Springer Theses

Recognizing Outstanding Ph.D. Research

For further volumes:
http://www.springer.com/series/8790

Aims and Scope

The series "Springer Theses" brings together a selection of the very best Ph.D. theses from around the world and across the physical sciences. Nominated and endorsed by two recognized specialists, each published volume has been selected for its scientific excellence and the high impact of its contents for the pertinent field of research. For greater accessibility to non-specialists, the published versions include an extended introduction, as well as a foreword by the student's supervisor explaining the special relevance of the work for the field. As a whole, the series will provide a valuable resource both for newcomers to the research fields described, and for other scientists seeking detailed background information on special questions. Finally, it provides an accredited documentation of the valuable contributions made by today's younger generation of scientists.

Theses are accepted into the series by invited nomination only and must fulfill all of the following criteria

- They must be written in good English.
- The topic should fall within the confines of Chemistry, Physics, Earth Sciences and related interdisciplinary fields such as Materials, Nanoscience, Chemical Engineering, Complex Systems and Biophysics.
- The work reported in the thesis must represent a significant scientific advance.
- If the thesis includes previously published material, permission to reproduce this must be gained from the respective copyright holder.
- They must have been examined and passed during the 12 months prior to nomination.
- Each thesis should include a foreword by the supervisor outlining the significance of its content.
- The theses should have a clearly defined structure including an introduction accessible to scientists not expert in that particular field.

Jonathan D. Pritchard

Cooperative Optical Non-Linearity in a Blockaded Rydberg Ensemble

Doctoral Thesis accepted by
Durham University, UK

 Springer

Author
Dr. Jonathan D. Pritchard
Durham University
Durham
UK

Supervisor
Prof. Charles Adams
Durham University
Durham
UK

ISSN 2190-5053
ISBN 978-3-642-29711-3
DOI 10.1007/978-3-642-29712-0
Springer Heidelberg New York Dordrecht London

ISSN 2190-5061 (electronic)
ISBN 978-3-642-29712-0 (eBook)

Library of Congress Control Number: 2012938200

Printed on acid-free paper

Springer is part of Springer Science+Business Media (www.springer.com)

This thesis is dedicated to Beth, Mum and Dad, for your love and continued support in all that I do

Supervisor's Foreword

In free space, light beams can pass through one another without interacting. Inside an optical medium there is a small interaction that is typically only observable using high intensity laser light. This interaction gives rise to the field of nonlinear optics which lies at the core of current information and communication technologies. However, the future technological process is dependent on our ability to control light at lower intensity, ideally at the quantum level of single photons. Single photon non-optics is the key that could open the door to fully deterministic quantum computing using light, and begin a new era of highly efficient information processing. However, single photon nonlinear optics has remained stubbornly outside the range of practical realization.

This thesis presents pioneering work aimed at developing a new approach to single photon nonlinear optics. The idea is to convert photons into excitations that are strongly interacting. This is achieved using an additional laser that couples an incoming photon into a highly excited atomic state known as a Rydberg state. An atom in a Rydberg state interacts strongly with neighboring Rydberg atoms. Consequently, the medium produces strong effective photon–photon interactions by mapping photons into Rydberg excitations and then back into photons. The dipole–dipole interactions between Rydberg atoms are larger than a single Rydberg excitation and hence a single photon can switch the optical response of many nearby atoms—a process known as dipole blockade. This cooperative response of many atoms acts like an amplifier, greatly enhancing the effect of each individual photon. This cooperative enhancement of the single photon nonlinearity means that strong photon–photon interactions are now accessible potentially providing the means to realize a fully deterministic all-optical quantum information processor.

The thesis describes the theoretical foundations of the cooperative nonlinearity beginning with the physics of Rydberg atoms, then atom–light interactions including dipole–dipole interactions, and cooperative phenomena. Subsequently, the experimental demonstration of the cooperative nonlinearity is described and finally the extension to the quantum regime is discussed. The work reported here has stimulated enormous interest worldwide and there are now many impressive new developments both theoretical and experimental appearing in the scientific literature. We are now entering an interesting time in the history of optics where strongly interacting photons become accessible.

Durham, UK, March 2012   Prof. Charles Adams

Acknowledgments

Over the last 3.5 years, I have had the privilege of working alongside a number of people without whom this thesis may have been significantly shorter! Most important thanks must go to my supervisor, Charles Adams, for giving me the opportunity to study here in Durham and for his guidance and infectious enthusiasm. Credit must also be given to Kev Weatherill for building a reliable, working experiment, his patience in showing me how to use it, and his continued friendship. Thanks to Alex Gauguet for hours spent performing painstaking lens alignment, and to Dan Maxwell, with whom I endured more knife-edging than I thought possible! Diolch yn fawr Ifan Hughes for always making time to answer my questions, and to Matt Jones for many useful discussions, and to all of the above for proof reading this thesis. I would also like to acknowledge fruitful collaboration with Klaus Mølmer, who showed us how to calculate the off-axis correlations from our model.

Research is never carried out in isolation, and I would like to thank all members of AtMol past and present for creating an open, friendly environment in which to study, including the many stimulating conversations on a Friday night in the Swan! Special mention must go to Tom Billam for his Fortran expertise, and to Mark Bason, Rich Abel, Ulrich Krohn, and especially James Millen for the banter that has made every day enjoyable.

Last, and by no means least, I want to thank my parents for their continued financial support and encouragement, and my wife Beth for three wonderful years of marriage and her enduring patience, particularly during the last 3 months—I am really looking forward to starting our new adventure together as parents.

Acknowledgments

Contents

Part II   Observations of Cooperativity

Chapter 1
Introduction

1.1 Introduction

The ability to manipulate the properties of light propagating through a medium was first discovered by Faraday [1] in 1846, who rotated the polarisation of light in lead glass using an external magnetic field. Similar observations were made by Kerr in 1875 [2] using a static electric field. However, with the advent of lasers in 1960 [3], the high optical intensities made it possible to modify the optical properties using the electric field of the light itself.

As light passes through a medium it is both attenuated and phase-shifted, providing control over the amplitude and polarisation. This optical response can be characterised in terms of the susceptibility χ, which is related to the refractive index of the medium by $n = \sqrt{1 + \chi}$. The susceptibility is a complex parameter, with the real part creating a dispersive phase shift and the the complex component leading to absorption of the light passing through the medium.

The non-linear response of the atom-light interaction is expressed in terms of a power expansion of the electric field E as [4]

$$\chi = \chi^{(1)} + \chi^{(2)} E + \chi^{(3)} E^2 + \cdots, \tag{1.1}$$

where $\chi^{(1)}$ represents the linear optical response and the higher-order terms describe non-linear optical processes, for example the Kerr effect corresponds to a $\chi^{(3)}$ process. One of the challenges in developing non-linear media relates to finding processes for which the attenuation associated with the linear susceptibility $\chi^{(1)}$ does not dominate over the higher-order effects.

J. D. Pritchard, *Cooperative Optical Non-Linearity in a Blockaded Rydberg Ensemble*,
Springer Theses, DOI: 10.1007/978-3-642-29712-0_1,
© Springer-Verlag Berlin Heidelberg 2012

1.1.1 Single-Photon Non-Linearities

Recently, attention has been focused on the development of non-linearities at the single-photon level for applications in quantum information processing (QIP) [5–7], where information is stored in a two-level quantum mechanical system known as a *qubit* [8]. QIP has the advantage of being able to use superposition states and entanglement between qubits to enable significant enhancement in the computation of classically 'hard' algorithms which rely on a brute force approach, such as searching through data [9] and prime number factorisation [10]. More important though is that as the information is represented by a quantum system, such a device could be used to directly simulate complex quantum many-body systems that cannot be modelled using digital computers.

Photons are ideal carriers of quantum information as they are robust against decoherence due to the incredibly weak interaction with the environment, and can be transmitted over long distances either in free-space or using optical fibres. Implementation of single qubit gates is trivially achieved using linear optics such as beam-splitters and polarisation optics, meaning photons fulfil most of the criteria for quantum computers laid out by D. DiVincenzo [11]. The key requirement for computation is the ability to perform deterministic two qubit gates to establish entanglement, such as a C-NOT or phase gate [8]. However, the downside of the weak interaction with the environment is that optical non-linearities at the single photon level are typically very weak [12, 13], making the implementation of two-photon gates challenging.

One solution proposed by Knill, Laflamme and Milburn (KLM) [14] is to use linear optics combined with ancillary photons to perform two-qubit gates probabilistically. This scheme has been implemented to perform a C-NOT gate [15–17], a π-phase gate [18] and a simple quantum circuit [19], however this approach has a number of drawbacks. The requirement for additional qubits, combined with the finite probability for success, makes the prospect of scaling this to performing computation unfavourable. The enhanced speed of the quantum algorithms is also negated by the need for many repetitions.

A more promising path to developing gates for quantum information is to exploit systems with a large single-photon non-linearity. Promising candidates include cavity QED [20, 21] or atom-light interactions in free space using electromagnetically induced transparency (EIT).

1.1.2 Electromagnetically Induced Transparency

EIT is the coherent phenomena arising from a three-level system coupled by a weak probe field and a strong coupling field. On resonance, this changes the medium from being optically thick to transparent for the probe transition [22]. This process can be understood from the formation of a *dark state*, which is a coherent superposition of the atomic levels of the system that no longer resonantly couples to the probe

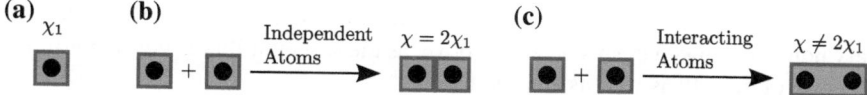

Fig. 1.1 Cooperative optical non-linearity. **a** A single atom has suscpeptibility χ_1. **b** For a pair of independent atoms, the susceptibility is $\chi = 2\chi_1$, and scales linearly for increasing atom number. **c** Dipole-dipole interactions modify the response of a pair of atoms such that $\chi \neq 2\chi_1$, and instead the susceptibility is a function of the separation between the atoms. This is a cooperative effect that cannot be solved using a single-atom picture

field [13] (*i.e.* $\chi^{(1)} \to 0$). This creates a narrow transmission window in the probe-only absorption feature, as first observed in a strontium vapour [23], resulting in a resonantly enhanced third-order non-linearity ($\chi^{(3)}$) [22].

Associated with this change in transmission is a concomitant modification of the dispersive properties of the medium, dramatically reducing the group velocity on resonance. This causes pulses to be slowed in the medium, as demonstrated using a sodium Bose–Einstein condensate (BEC) to obtain $v_g = 17\,\mathrm{ms}^{-1}$ [24], which is the largest measured Kerr non-linearity in an atomic system. Slow light is also possible in a room temperature vapour, enabling propagation at speeds of $v_g = 8\,\mathrm{ms}^{-1}$ [25].

In addition to slowing light, pulses can be halted and stored for up to 1 ms [26, 27] by turning off the coupling laser whilst the pulse is propagating through the medium. The ability to coherently convert photons into excitations in the medium [28] has lead to significant progress in development of quantum memories for photonic qubits [7, 29]. EIT has since been used to both generate, and store, single photon pulses to perform quantum communication between two remote quantum memories [30, 31].

The large non-linearity associated with EIT can be further enhanced using a four-level system with an additional laser field. This results in a giant Kerr non-linearity [5] capable of performing cross-phase modulation with a pair of photons, which was originally suggested as a non-linearity suitable for performing a conditional phase gate. However, subsequent work has shown that Kerr-type non-linearities cannot be used to obtain high-fidelity quantum gates due to distortion of the photons [32, 33].

The quest to achieve quantum gates therefore requires a novel non-linear optical process. A recent proposal suggests use of a spin-wave interaction in a BEC [34]. Below the effects of dipole–dipole interactions are considered.

1.1.3 Cooperative Effects Due to Dipole–Dipole Interactions

Typically, non-linear media can be understood as isolated, non-interacting quantum systems which are driven by optical fields [4], such as an atom. From the properties of a single atom, the optical properties of the medium can be understood from linearly scaling the system proportional to the number of atoms. However, if dipole–dipole interactions between the atoms are introduced, the properties of each atom now depend on the presence of the neighbouring atoms, resulting in cooperative phenomena, illustrated in Fig. 1.1. This is fundamentally different from the ordinary non-linear mechanisms detailed above, as the susceptibility is now a non-linear function of density as well as electric field.

The best example of cooperative behaviour due to dipole–dipole interactions is superradiance [35]. For an ensemble of $\mathcal{N}$-atoms initially in an excited state, the interaction of each atom with the dipole-field of all the surrounding atoms causes their dipoles to become phased. This leads to rapid decay on timescales much faster than the natural lifetime of the excited state, emitting an intense pulse of radiation [36].

Dipole–dipole interactions are only observable for samples with an interatomic separation $R < \lambda$, where λ is the wavelength of the transition. This corresponds to densities of order $10^{15}\,\mathrm{cm}^{-3}$ for optical transitions, at which densities the collisional dephasing in thermal samples precludes observation of superradiance without use of a BEC [37]. Consequently, a cooperative optical non-linearity has only been previously observed in an up-conversion process, requiring very high optical intensity [38].

1.1.4 Rydberg Atoms

The difficulty in achieving $R < \lambda$ can be overcome by using states coupled by microwave transitions, such as a Rydberg state [39]. Rydberg atoms represent highly excited states of the valence electron, which are relatively long-lived states with large orbital radii. Their large radius gives the Rydberg states a very large dipole moment, resulting in very strong long-range dipole–dipole interactions between atoms that shift the energy of the multiply excited Rydberg states [39]. When the energy shift exceeds the linewidth of the excitation laser, only a single atom can be excited to the Rydberg state, an effect known as *dipole blockade* [40]. This enables deterministic creation of a single excitation for atoms confined within around 5 μm, making Rydberg atoms ideally suited to studies of quantum many-body physics and quantum information processing.

The controllable interactions of the Rydberg states have been studied in a variety of regimes, demonstrating resonant energy transfer [41–44] and mechanical effects of dipole–dipole interactions, namely ionisation due to the attractive or repulsive potentials [45, 46]. Important steps towards exploiting the strong interactions for quantum information were the observation of coherent excitation of the Rydberg states [47–49] and the demonstration of dipole blockade [50–59].

A number of theoretical proposals exist to realise quantum gates [40, 60–62] and quantum simulators [63] using dipole blockade of the Rydberg states. Recently, the conditional entanglement [64] and a C-NOT gate [65] have been demonstrated for a pair of atoms separated by around 3 μm. Other interesting applications of the dipolar interactions include the creation of strongly correlated atomic states by weakly dressing ground-state atoms with Rydberg character [66, 67], or novel phase transitions from resonant coupling of the Rydberg states [68, 69].

As well as creating correlated atomic states, blockade can be used to create single photons through four-wave mixing [70], or collective emission of the shared excitation [71]. Generation of more complex quantum states of light is possible using a multi-level atomic system [72], or through photon subtraction [73].

1.1.5 Rydberg EIT

The properties of the Rydberg states can be mapped onto optical transitions using EIT. The original proposal by Friedler et al. to realise a photonic phase gate using Rydberg atoms relied on using EIT to create weakly interacting dark states that counter-propagate through an atomic vapour [74]. This treated the photons as being 1D, leading to an updated proposal of entanglement by applying this scheme in a hollow core fibre filled with an atomic sample [75]. One of the challenges with this approach is to overcome the strong interaction between the Rydberg atoms and the surface of the fibre [76].

Pioneering studies of Rydberg EIT were performed in a thermal vapour [77] to demonstrate EIT as a coherent, non-destructive probe of the Rydberg energy levels in contrast to detection using ionisation [39]. This also demonstrated EIT as a method of performing spectroscopy of the Rydberg states, which was also used in an atomic beam to measure isotope shifts in Sr [78]. Combining EIT with the techniques of FM spectroscopy, an error signal suitable for stabilising the frequency of the Rydberg excitation laser can be generated [79]. This work was of fundamental importance to the results in this thesis, as the ability to actively stabilise the frequency of the two-photon resonance enabled high resolution spectroscopy to be performed, achieving narrow sub-MHz resonances in a single measurement [80].

The Rydberg character of the EIT dark state has been exploited to control the propagation of a probe beam through the cell using external electric fields, demonstrating optical switching [81] and a giant electro-optic effect 10^6 times larger than for typical Kerr media due to the large polarisability of the Rydberg states. It has also been used for applications in electrometry [82, 83]. However, for thermal samples the effect of dipole–dipole interactions have yet to be observed due Doppler broadening reducing the size of the blockade [76].

Studies of dipole–dipole interaction effects in EIT of a cold atomic gas form the basis of this thesis, with initial experiments on low principal quantum number states displaying cooperative behaviour consistent with superradiance [80]. Complementary work revealed dipole–dipole interactions create a dephasing of the EIT [84], however it will be shown that for highly excited Rydberg states the dipole blockade mechanism can be used to obtain a cooperative optical non-linearity [85, 86]. This thesis describes the characterisation of the non-linearity, and progress towards realising this effect at the single photon level using a single blockade volume.

1.2 Thesis Layout

The document is separated into four parts—part I explores the properties of the Rydberg states and how the strong interactions can be mapped onto an optical field to obtain a novel optical non-linearity. Part II details the experimental setup and observation of cooperative effects through EIT spectroscopy of an ultra-cold atom cloud. Part III extends these ideas to consider the non-linearity at the single-photon level. Finally, part IV draws these results together to consider future areas of study. The chapter breakdown is as follows;

Part I: Rydberg Atom-Light Interactions

- Chapter 2 describes the properties of Rydberg atoms, detailing the calculation of dipole matrix elements and their application in computing Stark maps and static polarisabilities of each state.
- Chapter 3 contains the theory of dipole–dipole interactions between pairs of Rydberg atoms, discussing the long- and short-range regimes and their effect on motional dynamics. The ability to tune the sign and strength of the interaction using an external field is also considered.
- Chapter 4 outlines the theory of atom-light interactions required to calculate the optical properties of a single atom and to understand electromagnetically induced transparency (EIT).
- Chapter 5 considers the cooperative phenomena arising due to dipole–dipole inter-actions, namely superradiance and dipole-blockade. Blockade is discussed in the context of EIT, and a few-atom model used to illustrate the resulting cooperative non-linearity.

Part II: Observations of Cooperativity

- Chapter 6 gives an account of the experimental setup used to perform EIT on a cold, dense atomic ensemble, including details of the data analysis procedure.
- Chapter 7 presents the results of these experiments, demonstrating a superradiant loss for low-n states, and characterising the cooperative optical non-linearity due to both attractive and repulsive interactions for states around $n \sim 60$.

Part III: Rydberg Atom Quantum Optics

- Chapter 8 discusses the blockade mechanism in the context of quantum optics. A model is developed to show that a single, optically thick blockaded ensemble can be used to create a train of highly correlated single photons from a coherent input state.
- Chapter 9 describes the design and construction of a new apparatus to trap atoms in a single blockade volume, enabling studies of the optical non-linearity at the single photon level.

Part IV: Conclusion

- Finally, chapter 10 summarises the important results and discusses future directions for this research.

1.3 Publications Arising from this Work

- J. D. Pritchard, C. S. Adams and K. Mølmer, *Correlated Photon Emission from Multiatom Rydberg Dark States,* Phys. Rev. Lett **108**, 043601 (2012). doi:10.1103/PhysRevLett.108.043601
- S. Sevinçli et al., *Quantum interference in interacting three-level Rydberg gases: Coherent Population Trapping and Electromagnetically-Induced Transparency,* J. Phys. B **44**, 184018 (2011). doi:10.1088/0953-4075/44/18/184018
- J. D. Pritchard, A. Gauguet, K. J. Weatherill, and C. S. Adams, *Optical nonlinearity in a dynamical Rydberg gas,* J. Phys. B **44**, 184019 (2011). doi:10.1088/0953-4075/44/18/184019
- M. Tanasittikosol, J. D. Pritchard, D. Maxwell, A. Gauguet, K. J. Weatherill, R. M. Potvliege, and C. S. Adams, *Microwave dressing of Rydberg dark states,* J. Phys. B **44**, 184020 (2011). doi:10.1088/0953-4075/44/18/184020
- J. D. Pritchard, D. Maxwell, A. Gauguet, K. J. Weatherill, M. P. A. Jones and C. S. Adams, *Cooperative atom-light interaction in a blockaded Rydberg ensemble,* Phys. Rev. Lett. **105**, 193603 (2010). doi:10.1103/PhysRevLett.105.193603
- R. P. Abel, A. K. Mohapatra, M. G. Bason, J. D. Pritchard, K. J. Weatherill, U. Raitzsch, and C. S. Adams, *Laser frequency stabilization to highly excited state transitions using electromagnetically induced transparency in a cascade system,* Appl. Phys. Lett. **94**, 071107 (2009). doi:10.1063/1.3086305
- K. J. Weatherill, J. D. Pritchard, P. F. Griffin, U. Dammalapati, C. S. Adams, and E. Riis, *A versatile and reliably re-usable ultra-high vacuum viewport,* Rev. Sci. Inst. **80**, 026105 (2009). doi:10.1063/1.3075547
- K. J. Weatherill, J. D. Pritchard, R. P. Abel, M. G. Bason, A. K. Mohapatra, and C. S. Adams, *Electromagnetically induced transparency of an interacting cold Rydberg ensemble,* J. Phys. B **41**, 201002 (2008). doi:10.1088/0953-4075/41/20/201002

References

1. M. Faraday, Experimental Researches in Electricity. Phil. Trans. R. Soc. Lond. **136**, 1 (1846)
2. J. Kerr, A new relation between electricity and light: dielectrified media birefringent. Phil. Mag. **50**, 337 (1875)
3. T.H. Maiman, Stimulated optical radiation in ruby. Nature **187**, 493 (1960)
4. R.W. Boyd, *Nonlinear Optics*, 3rd edn. (Academic Press, San Diego, 2008)
5. H. Schmidt, A. Imamoglu, Giant Kerr nonlinearities obtained by electromagnetically induced transparency. Opt. Lett. **21**(23), 1936 (1996)
6. M.D. Lukin, A. Imamoglu, Nonlinear optics and quantum entanglement of ultraslow single photons. Phys. Rev. Lett. **84**(7), 1419 (2000)
7. M.D. Lukin, Colloquium: trapping and manipulating photon states in atomic ensembles. Rev. Mod. Phys. **75**(2), 457 (2003)
8. M.A. Nielsen, I.L. Chuang, *Quantum Computation and Quantum Information* (CUP, Cambridge, 2005)

9. L.K. Grover, Quantum mechanics helps in searching for a needle in a haystack. Phys. Rev. Lett. **79**(2), 325 (1997)
10. P.W. Shor, Polynomial—time algorithms for prime factorization and discrete logarithms on a quantum computer. SIAM J. Comput. **26**, 1484 (1997)
11. D.P. DiVincenzo, D. Bacon, J. Kempe, G. Burkard, K.B. Whaley, Universal quantum computation with the exchange interaction. Nature **408**, 339 (2000)
12. N. Matsuda, R. Shimizu, Y. Mitsumori, H. Kosaka, K. Edamatsu, Observation of optical-fibre Kerr nonlinearity at the single-photon level. Nature Photon. **3**, 95 (2009)
13. M. Fleischhauer, A. Imamoglu, J. Marangos, Electromagnetically induced transparency: optics in coherent media. Rev. Mod. Phys. **77**, 633 (2005)
14. E. Knill, R. Laflamme, G.J. Milburn, A scheme for efficient quantum computation with linear optics. Nature **409**, 46 (2001)
15. T.B. Pittman, B.C. Jacobs, J.D. Franson, Demonstration of nondeterministic quantum logic operations using linear optical elements. Phys. Rev. Lett. **88**(25), 257902 (2002)
16. L. O'Brien, G.J. Pryde, A.G. White, T.C. Ralph, D. Branning, Demonstration of an all—optical quantum controlled—NOT gate. Nature **426**, 264 (2003)
17. S. Gasparoni, J.-W. Pan, P. Walther, T. Rudolph, A. Zeilinger, Realization of a photonic controlled—NOT gate sufficient for quantum computation. Phys. Rev. Lett. **93**(2), 020504 (2004)
18. K. Sanaka, T. Jennewein, J.-W. Pan, K. Resch, A. Zeilinger, Experimental nonlinear sign shift for linear optics quantum computation. Phys. Rev. Lett. **92**(1), 017902 (2004)
19. T.B. Pittman, B.C. Jacobs, J.D. Franson, Experimental demonstration of a quantum circuit using linear optics gates. Phys. Rev. A **71**(3), 032307 (2005)
20. Q.A. Turchette, C.J. Hood, W. Lange, H. Mabuchi, H.J. Kimble, Measurement of conditional phase shifts for quantum logic. Phys. Rev. Lett. **75**(25), 4710 (1995)
21. A. Rauschenbeutel, P. Bertet, S. Osnaghi, G. Nogues, M. Brune, J.M. Raimond, S. Haroche, Controlled entanglement of two field modes in a cavity quantum electrodynamics experiment. Phys. Rev. A **64**(5), 050301 (2001)
22. S.E. Harris, J.E. Field, A. Imamoglu, Nonlinear optical processes using electromagnetically induced transparency. Phys. Rev. Lett. **64**(10), 1107 (1990)
23. K.-J. Boller, A. Imamoglu, S.E. Harris, Observation of electromagnetically induced transparency. Phys. Rev. Lett. **66**(20), 2593 (1991)
24. L.V. Hau, S.E. Harris, Z. Dutton, C.H. Behroozi, Light Speed reduction to 17 metres per second in an ultracold atomic gas. Nature **397**, 594 (1999)
25. D. Budker, D.F. Kimball, S.M. Rochester, V.V. Yashchuk, Nonlinear magneto-optics and reduced group velocity of light in atomic vapor with slow ground state relaxation. Phys. Rev. Lett. **83**(9), 1767 (1999)
26. C. Liu, Z. Dutton, C.H. Behroozi, L.V. Hau, Observation of coherent optical information storage in an atomic medium using halted light pulses. Nature **409**, 490 (2001)
27. D.F. Phillips, A. Fleischhauer, A. Mair, R.L. Walsworth, M.D. Lukin, Storage of light in atomic vapor. Phys. Rev. Lett. **86**(5), 783 (2001)
28. M. Fleischhauer, M.D. Lukin, Dark-state polaritons in electromagnetically induced transparency. Phys. Rev. Lett. **84**(22), 5094 (2000)
29. C. Simon et al., Quantum memories. Eur. Phys. J. D. **58**, 1 (2010)
30. M.D. Eisaman, A. André, F. Massou, M. Fleischhauer, A.S. Zibrov, M.D. Lukin, Electromagnetically induced transparency with tunable single-photon pulses. Nature **438**, 837 (2005)
31. T. Chanelière, D.N. Matsukevich, S.D. Jenkins, S.-Y. Lan, T.A.B. Kennedy, A. Kuzmich, Storage and retrieval of single photons transmitted between remote quantum memories. Nature **438**, 833 (2005)
32. J.H. Shapiro, Single-photon Kerr nonlinearities do not help quantum computation. Phys. Rev. A **73**(6), 062305 (2006)
33. J. Gea-Banacloche, Impossibility of large phase shifts via the giant Kerr effect with single-photon wave packets. Phys. Rev. A **81**(4), 043823 (2010)

34. A.V. Gorshkov, J. Otterbach, E. Demler, M. Fleischhauer, M.D. Lukin, Photonic phase gate via an exchange of fermionic spin waves in a spin chain. Phys. Rev. Lett. **105**(6), 060502 (2010)
35. R.H. Dicke, Coherence in spontaneous radiation processes. Phys. Rev. **93**(1), 99 (1954)
36. M. Gross, S. Haroche, Superradiance: an essay on the theory of collective spontaneous emission. Phys. Rep. **93**(5), 301 (1982)
37. S. Inouye, A.P. Chikkatur, D.M. Stamper-Kurn, J. Stenger, D.E. Pritchard, W. Ketterle, Super-radiant Rayleigh scattering from a Bose–Einstein condensate. Science **285**(5427), 571 (1999)
38. M.P. Hehlen, H.U. Güdel, Q. Shu, J. Rai, S. Rai, S.C. Rand, Cooperative bistability in dense, excited atomic systems. Phys. Rev. Lett. **73**(8), 1103 (1994)
39. T.F. Gallagher, *Rydberg Atoms* (CUP, New York, 2005)
40. M.D. Lukin, M. Fleischhauer, R. Cote, L.M. Duan, D. Jaksch, J.I. Cirac, P. Zoller, Dipole blockade and quantum information processing in mesoscopic atomic ensembles. Phys. Rev. Lett. **87**(3), 037901 (2001)
41. W.R. Anderson, J.R. Veale, T.F. Gallagher, Resonant dipole–dipole energy transfer in a nearly Frozen Rydberg gas. Phys. Rev. Lett. **80**(2), 249 (1998)
42. M. Mudrich, N. Zahzam, T. Vogt, D. Comparat, P. Pillet, Back and forth transfer and coherent coupling in a cold Rydberg dipole gas. Phys. Rev. Lett. **95**(23), 233002 (2005)
43. S. Westermann, T. Amthor, A.L. de Oliveira, J. Deiglmayr, M. Reetz-Lamour, M. Weidemuller, Dynamics of resonant energy transfer in a cold Rydberg gas. Eur. Phys. J. D. **40**, 37 (2006)
44. J.A. Petrus, P. Bohlouli-Zanjani, J.D.D. Martin, ac electric-field-induced resonant energy transfer between cold Rydberg atoms. J. Phys. B **41**, 245001 (2008)
45. T. Amthor, M. Reetz-Lamour, S. Westermann, J. Denskat, M. Weidemüller, Mechanical effect of van der Waals interactions observed in real time in an ultracold Rydberg gas. Phys. Rev. Lett. **98**(2), 023004 (2007)
46. T. Amthor, M. Reetz-Lamour, C. Giese, M. Weidemüller, Modeling many-particle mechanical effects of an interacting Rydberg gas. Phys. Rev. A **76**(5), 054702 (2007)
47. J. Dieglmayr, M. Reetz-Lamour, T. Amthor, S. Westermann, A.L. de Oliveira, M. Weidemüller, Coherent excitation of Rydberg atoms in an ultracold gas. Opt. Comm. **264**, 293 (2006)
48. M. Reetz-Lamour, J. Deiglmayr, T. Amthor, M. Weidemüller, Rabi oscillations between ground and Rydberg states and van der Waals blockade in a mesoscopic frozen Rydberg gas. New J. Phys. **10**(4), 045026 (2008)
49. T.A. Johnson, E. Urban, T. Henage, L. Isenhower, D.D. Yavuz, T.G. Walker, M. Saffman, Rabi oscillations between ground and Rydberg states with dipole–dipole atomic interactions. Phys. Rev. Lett. **100**(11), 113003 (2008)
50. D. Tong, S.M. Farooqi, J. Stanojevic, S. Krishnan, Y.P. Zhang, R. Côté, E.E. Eyler, P.L. Gould, Local blockade of Rydberg excitation in an ultracold gas. Phys. Rev. Lett. **93**(6), 063001 (2004)
51. K. Singer, M. Reetz-Lamour, T. Amthor, L.G. Marcassa, M. Weidemüller, Suppression of excitation and spectral broadening induced by interactions in a cold gas of Rydberg atoms. Phys. Rev. Lett. **93**(16), 163001 (2004)
52. K. Afrousheh, P. Bohlouli-Zanjani, D. Vagale, A. Mugford, M. Fedorov, J.D.D. Martin, Spectroscopic observation of resonant electric dipole–dipole interactions between cold Rydberg atoms. Phys. Rev. Lett. **93**(23), 233001 (2004)
53. T. Cubel Liebisch, A. Reinhard, P.R. Berman, G. Raithel, Atom counting statistics in ensembles of interacting Rydberg atoms. Phys. Rev. Lett. **95**(25), 253002 (2005)
54. T. Vogt, M. Viteau, J. Zhao, A. Chotia, D. Comparat, P. Pillet, Dipole blockade at Förster resonances in high resolution laser excitation of Rydberg states of cesium atoms. Phys. Rev. Lett. **97**(8), 083003 (2006)
55. T. Vogt, M. Viteau, A. Chotia, J. Zhao, D. Comparat, P. Pillet, Electric-field induced dipole blockade with Rydberg atoms. Phys. Rev. Lett. **99**(7), 073002 (2007)
56. C.S.E. van Ditzhuijzen, A.F. Koenderink, J.V. Hernández, F. Robicheaux, L.D. Noordam, H.B. van Linden, van den Heuvell, Spatially resolved observation of dipole–dipole interaction between Rydberg atoms. Phys. Rev. Lett. **100**(24), 243201 (2008)
57. R. Heidemann, U. Raitzsch, V. Bendkowsky, B. Butscher, R. Low, L. Santos, T. Pfau, Evidence for coherent collective Rydberg excitation in the strong blockade regime. Phys. Rev. Lett. **99**(16), 163601 (2007)

58. E. Urban, T.A. Johnson, T. Henage, L. Isenhower, D.D. Yavuz, T.G. Walker, M. Saffman, Observation of Rydberg blockade between two atoms. Nature Phys. **5**, 110 (2009)

59. A. Gaëtan, Y. Miroshnychenko, T. Wilk, A. Chotia, M. Viteau, D. Comparat, P. Pillet, A. Browaeys, P. Grangier, Observation of collective excitation of two individual atoms in the Rydberg blockade regime. Nature Phys. **5**, 115 (2009)

60. D. Jaksch, J.I. Cirac, P. Zoller, Fast quantum gates for neutral atoms. Phys. Rev. Lett. **85**(10), 2208 (2000)

61. D. Møller, L.B. Madsen, K. Mølmer, Quantum gates and multiparticle entanglement by Rydberg excitation blockade and adiabatic passage. Phys. Rev. Lett. **100**(17), 170504 (2008)

62. M. Müller, I. Lesanovsky, H. Weimer, H.P. Büchler, P. Zoller, Mesoscopic Rydberg gate based on electromagnetically induced transparency. Phys. Rev. Lett. **102**(17), 170502 (2009)

63. H. Weimer, M. Müller, I. Lesanovsky, P. Zoller, H.P. Büchler, A Rydberg quantum simulator. Nature Phys. **6**, 382 (2010)

64. T. Wilk, A. Gaëtan, C. Evellin, J. Wolters, Y. Miroshnychenko, P. Grangier, A. Browaeys, Entanglement of two individual neutral atoms using Rydberg blockade. Phys. Rev. Lett. **104**(1), 010502 (2010)

65. L. Isenhower, E. Urban, X.L. Zhang, A.T. Gill, T. Henage, T.A. Johnson, T.G. Walker, M. Saffman, Demonstration of a neutral atom controlled-NOT quantum gate. Phys. Rev. Lett. **104**(1), 010503 (2010)

66. G. Pupillo, A. Micheli, M. Boninsegni, I. Lesanovsky, P. Zoller, Strongly correlated gases of Rydberg-dressed atoms: quantum and classical dynamics. Phys. Rev. Lett. **104**(22), 223002 (2010)

67. T. Pohl, E. Demler, M.D. Lukin, Dynamical crystallization in the dipole blockade of ultracold atoms. Phys. Rev. Lett. **104**(4), 043002 (2010)

68. H. Weimer, R. Löw, T. Pfau, H.P. Büchler, Quantum critical behavior in strongly interacting Rydberg gases. Phys. Rev. Lett. **101**(25), 250601 (2008)

69. F. Cinti, P. Jain, M. Boninsegni, A. Micheli, P. Zoller, G. Pupillo, Supersolid droplet crystal in a dipole-blockaded gas. Phys. Rev. Lett. **105**(13), 135301 (2010)

70. M. Saffman, T.G. Walker, Creating single-atom and single-photon sources from entangled atomic ensembles. Phys. Rev. A **66**, 065403 (2002)

71. L.H. Pedersen, K. Mølmer, Few qubit atom—light interfaces with collective encoding. Phys. Rev. A **79**(1), 012320 (2009)

72. A.E.B. Nielsen, K. Mølmer, Deterministic multi-mode photonic device for quantum information processing. Phys. Rev. A **81**, 043822 (2010)

73. J. Honer, R. Löw, H. Weimer, T. Pfau, H.P. Büchler, Artificial atoms can do more than atoms: deterministic single photon subtraction from arbitrary light fields. Phys. Rev. Lett. **107**, 093601 (2011)

74. I. Friedler, D. Petrosyan, M. Fleischhauer, G. Kurizki, Long-range interactions and entanglement of slow single-photon pulses. Phys. Rev. A **72**, 043803 (2005)

75. E. Shahmoon, G. Kurizki, M. Fleischhauer, D. Petrosyan, Strongly interacting photons in hollow-core waveguides. Phys. Rev. A **83**, 033806 (2011)

76. H. Kübler, J.P. Shaffer, T. Baluktsian, R. Löw, T. Pfau, Coherent excitation of Rydberg atoms in micrometre-sized atomic vapour cells. Nature Photon. **4**, 112 (2010)

77. A.K. Mohapatra, T.R. Jackson, C.S. Adams, Coherent optical detection of highly excited Rydberg states using electromagnetically induced transparency. Phys. Rev. Lett. **98**(11), 113003 (2007)

78. S. Mauger, J. Millen, M.P.A. Jones, Spectroscopy of strontium Rydberg states using electromagnetically induced transparency. J. Phys. B **40**(22), F319 (2007)

79. R.P. Abel, A.K. Mohapatra, M.G. Bason, J.D. Pritchard, K.J. Weatherill, U. Raitzsch, C.S. Adams, Laser frequency stabilization to excited state transitions using electromagnetically induced transparency in a cascade system. Appl. Phys. Lett. **94**(7), 071107 (2009)

80. K.J. Weatherill, J.D. Pritchard, R.P. Abel, M.G. Bason, A.K. Mohapatra, C.S. Adams, Electromagnetically induced transparency of an interacting cold Rydberg ensemble. J. Phys. B **41**(20), 201002 (2008)

81. M.G. Bason, A.K. Mohapatra, K.J. Weatherill, C.S. Adams, Electro—optic control of atom—light interactions using Rydberg dark-state polaritons. Phys. Rev. A **77**, 032305 (2008)
82. M.G. Bason, M. Tanasittikosol, A. Sargsyan, A.K. Mohapatra, D. Sarkisyan, R.M. Potvliege, C.S. Adams, Enhanced electric field sensitivity of rf-dressed Rydberg dark states. New J. Phys. **12**, 065015 (2010)
83. A. Tauschinsky, R.M.T. Thijssen, S. Whitlock, H.B. van Linden van den Heuvell, R.J.C. Spreeuw, Spatially resolved excitation of Rydberg atoms and surface effects on an atom chip. Phys. Rev. A **81**, 063411 (2010)
84. U. Raitzsch, R. Heidemann, H. Weimer, V. Bendkowsky, B. Butscher, P. Kollmann, R. Löw, H.P. Büchler, T. Pfau, Investigation of dephasing rates in an interacting Rydberg gas. New J. Phys. **11**, 055014 (2009)
85. J.D. Pritchard, D. Maxwell, A. Gauguet, K.J. Weatherill, M.P.A. Jones, C.S. Adams, Cooperative atom—light interaction in a blockaded Rydberg ensemble. Phys. Rev. Lett. **105**(19), 193603 (2010)
86. C. Ates, S. Sevilçi, T. Pohl, Electromagnetically induced transparency in strongly interacting Rydberg gases. Phys. Rev. A **83**(4), 041802 (2011)

Part I
Rydberg Atom-Light Interactions

Chapter 2
Rydberg Atoms

The Rydberg series was originally identified in the spectral lines of atomic hydrogen, where the binding energy W was found empirically to be related to the formula [1]

$$W = -\frac{\text{Ry}}{n^2},\tag{2.1}$$

where Ry was a constant and n an integer. The theoretical underpinning for this scaling arrived with the Bohr model of the atom in 1913 [2], from which the Rydberg constant Ry could be derived in terms of fundamental constants

$$\text{Ry} = \frac{Z^2 e^4 m_e}{16\pi^2 \epsilon_0^2 \hbar^2},\tag{2.2}$$

and n understood as the principal quantum number. From the Bohr model it was also possible to derive scaling laws for the atomic properties in terms of n, which were later verified from the full quantum mechanical treatment of Schrödinger in 1926 [3]. Table 2.1 summarises the scalings of the atomic properties for the low-ℓ Rydberg states. The most important property of the Rydberg states is the large orbital radius, and hence dipole moment, $\propto n^2$. The consequence of the incredibly large dipole moment is an exaggerated response to external fields and the ability to observe dipole-dipole interactions between atoms on the μm scale. Combining this with the relatively long lifetimes, Rydberg atoms are well suited to applications in coherent quantum gates [4].

2.1 Alkali Metal Atom Rydberg States

Alkali metal atoms are similar to hydrogen, with a single valence electron orbiting a positively charged core which gives a $-1/r$ Coulomb potential at long range. However, the nucleus is surrounded by closed electron shells which screen the nuclear

J. D. Pritchard, *Cooperative Optical Non-Linearity in a Blockaded Rydberg Ensemble*,
Springer Theses, DOI: 10.1007/978-3-642-29712-0_2,
© Springer-Verlag Berlin Heidelberg 2012

Table 2.1 Scaling laws for properties of the Rydberg states [5]

Property	n-scaling
Binding energy W	n^{-2}
Orbital radius	n^2
Energy difference of adjacent n states Δ	n^{-3}
Radiative lifetime τ	n^{-3}

charge, giving the core a finite size. For the low orbital angular momentum states with $\ell \leq 3$, the electron orbit is extremely elliptic and can penetrate the closed electron shells. This exposes the valence electron to the unscreened nuclear charge, causing the core potential to deviate from the Coulombic potential at short range. The inner electrons can also be polarised by the valence electron. These two interactions with the core combine to increase the binding energy of the low-ℓ Rydberg states relative to the equivalent hydrogenic states. This difference in binding energy is parameterised using the quantum defects $\delta_{n\ell j}$

$$W = -\frac{\text{Ry}}{(n - \delta_{n\ell j})^2}, \tag{2.3}$$

where for rubidium the Rydberg constant is Ry $= 109736.605 \, \text{cm}^{-1}$ [6]. The properties of the alkali metal Rydberg states are thus determined from the effective principal quantum number $n^* = n - \delta_{n\ell j}$.

The value of the quantum defects depends on the quantum numbers for the Rydberg state of interest, where the S states have the largest defects as they have a significant core penetration. The quantum defects are determined empirically from spectroscopic measurements and can be calculated using

$$\delta_{n\ell j} = \delta_0 + \frac{\delta_2}{(n - \delta_0)^2} + \frac{\delta_4}{(n - \delta_0)^4} + \dots, \tag{2.4}$$

where $\delta_0, \delta_2 \dots$ are dependent upon ℓ and j. For rubidium, these have been measured on a cloud of cold atoms by the group of T. F. Gallagher and can be found in Ref. [7] for the S, P and D states and Ref. [8] for the F states. For $\ell > 3$ the quantum defects are zero, and the core potential is purely Coulombic. These are referred to as the hydrogenic states, which are degenerate for a given n.

2.2 Rydberg Atom Wavefunctions

The wavefunction for the valence electron is described by the Schrödinger equation, given in atomic units (a.u.) as

$$\left[-\frac{1}{2\mu}\nabla^2 + V(r)\right]\psi(r, \theta, \phi) = W\psi(r, \theta, \phi), \tag{2.5}$$

where μ is the reduced mass of the electron, r is the radial coordinate and $V(r)$ is the core potential. Since $V(r)$ has no angular dependence, the wavefunction is separable, giving $\psi(r, \theta, \phi) = R(r)Y_\ell^{m_\ell}(\theta, \phi)$, where $Y_\ell^{m_\ell}(\theta, \phi)$ is a spherical harmonic dependent upon the orbital angular momentum ℓ of the Rydberg state. Inserting this into Eq. 2.5 gives the equation for the radial wavefunction of the electron

$$\left[-\frac{1}{2\mu}\left(\frac{d^2}{dr^2} + \frac{2}{r}\frac{d}{dr}\right) + \frac{\ell(\ell+1)}{2\mu r^2} + V(r)\right]R(r) = WR(r). \tag{2.6}$$

Model Potential $V_C(r)$

To calculate the radial wavefunctions of the alkali metal atoms, it is necessary to use an ℓ-dependent core potential $V_C(r)$ to include the effects of core penetration and polarisation. This is done using a model potential given by [9]

$$V_C(r) = -\frac{Z_{n\ell}(r)}{r} - \frac{\alpha_c}{2r^4}(1 - e^{-(r/r_c)^6}). \tag{2.7}$$

The first term describes the Coulomb potential for a radial charge $Z_{n\ell}(r)$ to account for core penetration, where radial charge is defined as

$$Z_{n\ell}(r) = 1 + (Z - 1)e^{-a_1 r} - r(a_3 + a_4 r)e^{-a_2 r}. \tag{2.8}$$

The second term in Eq. 2.7 describes the long range potential of the induced core polarisation on the valence electron. The strength of this effect is determined by the core polarisability α_c, which increases with the number of electrons in the core.

Values for the parameters a_{1-4}, r_c and α_c are taken from Marinescu et al. [9], where the authors fit this model for the core potential to the measured energies of the Rydberg states for each ℓ-series of the alkali metals.

In addition to the core potential, the spin-orbit potential $V_{SO}(r)$ which causes the fine-structure splitting must also be included as [10]

$$V_{SO}(r) = \frac{\alpha^2}{2r^3}L \cdot S, \tag{2.9}$$

where α is the fine-structure constant and

$$L \cdot S = \frac{j(j+1) - \ell(\ell+1) - s(s+1)}{2}. \tag{2.10}$$

The total potential is thus $V(r) = V_c(r) + V_{SO}(r)$.

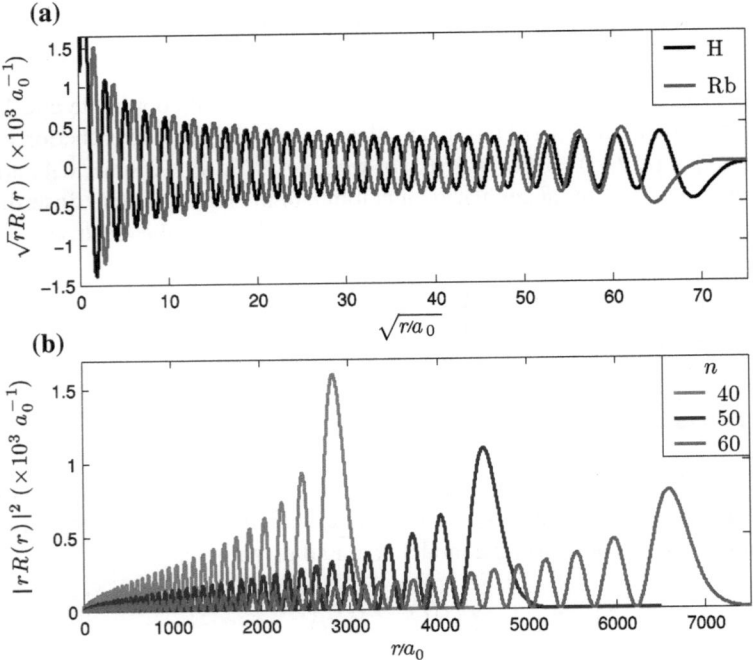

Fig. 2.1 Rydberg atom radial wavefunctions. **a** $50S_{1/2}$ radial wavefunction for rubidium and hydrogen. **b** Radial probability density for $nD_{5/2}$ states, illustrating the scaling of the radial wavefunction with n^{*2}

Numerical Integration

Using this model potential, the radial wavefunctions can be calculated by numerically integrating the radial Schrödinger equation of Eq. 2.6. This is simplified by performing a transformation to integrate the function $X(r) = R(r)r^{3/4}$ in terms of the scaled co-ordinate $x = \sqrt{r}$ [11]. This transformation converts Eq. 2.6 to a form solved efficiently using the Numerov algorithm [12, 13], whilst using the coordinate x gives an approximately constant number of points across each period of oscillation in the wavefunction. It is necessary to truncate the range of integration as at short range the model becomes unphysical and diverges, whilst at long range the wavefunction decays to zero. Following Ref. [14], the limits of integration are set to use an inner radius of $r_i = \sqrt[3]{\alpha_c}$, and an outer radius of $r_o = 2n(n + 15)$ which is much larger than the classical turning point of the wavefunction. To minimise errors introduced by the approximate model potential at short range, the integration is performed inwards, starting at r_o.

Figure 2.1a shows the calculated wavefunctions of the $50S_{1/2}$ states for hydrogen and rubidium as a function of the scaled coordinate. Comparing the two wavefunctions, the rubidium wavefunction is shifted to shorter radius relative to the hydrogen wavefunction due to the increased binding energy from the interaction with the core.

In (b) the electron probability density is plotted for the $nD_{5/2}$ states, illustrating the large orbital radii of the Rydberg states.

2.3 Dipole Matrix Elements

Transitions between atomic states primarily occur due to coupling with the electric dipole moment $\mu = er$ of the valence electron, which is a factor of $(\alpha/2)^2$ stronger than the magnetic dipole coupling [15]. The strength of the coupling between states $|n\ell m_\ell\rangle$ and $|n'\ell'm'_\ell\rangle$ is given by the dipole matrix element $\langle n\ell m_\ell|\mu|n'\ell'm'_\ell\rangle$, which is dependent upon the overlap of the wavefunctions with the electric dipole moment. From knowledge of the dipole matrix elements, it is possible to calculate transition probabilities, radiative lifetimes and many other properties of the atomic states [10].

The dipole operator is $\mu = er \cdot \hat{e}$, where $\hat{e}$ is the electric field polarisation unit vector. Transforming into the spherical basis, the dipole operator can be decomposed into the operators μ_q, with $q = \{-1, 0, +1\}$ corresponding to $\{\sigma^+, \pi, \sigma^-\}$ transitions, given by

$$\mu_{-1} = \frac{1}{\sqrt{2}}(\mu_x - i\mu_y), \tag{2.11a}$$

$$\mu_0 = \mu_z, \tag{2.11b}$$

$$\mu_{+1} = \frac{1}{\sqrt{2}}(\mu_x + i\mu_y). \tag{2.11c}$$

These operators are related to the spherical harmonics by $\mu_q = er\sqrt{4\pi/3}Y_1^q$ (θ, ϕ), which form a set of rank-1 irreducible tensors. As a result the Wigner–Eckart theorem can be used to separate dipole matrix element into an angular coupling and a reduced matrix element $\langle \ell||er||\ell'\rangle$ which depends only on ℓ and the radial wavefunctions [16]

$$\langle n\ell m_\ell|\mu_q|n'\ell'm'_\ell\rangle = (-1)^{\ell-m_\ell}\begin{pmatrix} \ell & 1 & \ell' \\ -m_\ell & q & m'_\ell \end{pmatrix}\langle\ell||\mu||\ell'\rangle, \tag{2.12}$$

where the brackets denote the Wigner-$3j$ symbol. Using the properties of the Wigner-$3j$ symbol, the selection rules of the electric dipole can be derived as $\Delta\ell = \pm 1$ and $\Delta m_\ell = 0, \pm 1$ corresponding to $\pi, \sigma^\pm$ transitions.

The reduced matrix element is defined as [17]

$$\langle\ell||\mu||\ell'\rangle = (-1)^\ell\sqrt{(2\ell + 1)(2\ell' + 1)}\begin{pmatrix} \ell & 1 & \ell' \\ 0 & 0 & 0 \end{pmatrix}\langle n\ell|er|n'\ell'\rangle, \tag{2.13}$$

where the radial matrix elements $\langle n\ell|er|n'\ell'\rangle$ represent the overlap integral between the radial wavefunctions and the dipole moment

$$\langle n\ell |er|n'\ell'\rangle = \int_{r_i}^{r_o} R_{n,\ell}(r)er\, R_{n,\ell'}(r)r^2\,dr, \qquad (2.14)$$

This can be evaluated by numerical integration over the wavefunctions calculated using the method described above.

2.3.1 Fine Structure Basis

The fine structure interaction V_{SO} breaks the degeneracy of the ℓ states, which split according to $j = \ell + s$. As the electric field only couples to the orbital angular momentum (ℓ) of the electron, it is therefore necessary to transform from the fine-structure basis into the uncoupled basis to evaluate the dipole matrix elements. Using the Wigner–Eckart theorem (Eq. 2.12), the matrix element can be expressed in terms of the reduced matrix element $\langle j||\mu||j'\rangle$. This is related to $\langle \ell||\mu||\ell'\rangle$ by [16]

$$\langle j||\mu||j'\rangle = (-1)^{\ell+s+j'+1}\delta_{s,s'}\sqrt{(2j+1)(2j'+1)}\begin{Bmatrix} j & 1 & j' \\ \ell' & s & \ell \end{Bmatrix}\langle \ell||\mu||\ell'\rangle, \quad (2.15)$$

where the braces denote a Wigner-$6j$ symbol. Combining these equations, the dipole matrix element in the fine-structure basis is

$$\langle n\ell j m_j|\mu_q|n'\ell'j'm_j'\rangle = (-1)^{j-m_j+s+j'+1}\sqrt{(2j+1)(2j'+1)(2\ell+1)(2\ell'+1)}$$

$$\times \begin{Bmatrix} j & 1 & j' \\ \ell' & s & \ell \end{Bmatrix}\begin{pmatrix} j & 1 & j' \\ -m_j & q & m_j' \end{pmatrix}\begin{pmatrix} \ell & 1 & \ell' \\ 0 & 0 & 0 \end{pmatrix}\langle n\ell j'|er|n'\ell'j'\rangle.$$

$$(2.16)$$

2.3.2 Hyperfine Structure Basis

The hyperfine interaction couples the angular momentum of the electron (j) and the nucleus (I), further lifting the degeneracy of the states which are split according to the total angular momentum $F = j + I$. As with the fine-structure splitting, the Wigner–Eckart theorem can be used to find the matrix elements in the hyperfine basis in terms of the reduced matrix element $\langle F||\mu||F'\rangle$, which can similarly be reduced to $\langle j||\mu||j'\rangle$.

For the Rydberg states the hyperfine splitting is typically small compared to the interaction with external fields e.g. $\nu_{hfs} \simeq 200\,kHz$ at $n = 60S_{1/2}$ [7]. The hyperfine splitting can therefore be neglected, treating Rydberg atoms in the fine-structure basis.

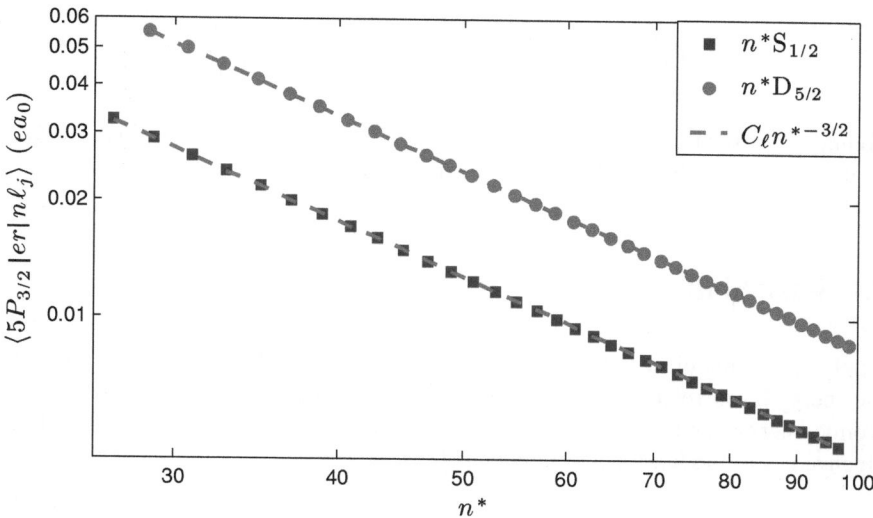

Fig. 2.2 Radial matrix elements for $5P_{3/2}$ to $nS_{1/2}$ or $nD_{5/2}$ transitions. The matrix elements scale as $n^{*-3/2}$

2.3.3 Rydberg Excitation Transition Strengths

In the experiments presented in this thesis, Rydberg states are excited by a two-photon transition in rubidium, using a laser at 780 nm to excite from the $5S_{1/2}$ ground-state to the $5P_{3/2}$ excited state, and a second laser at 480 nm to couple from $5P_{3/2}$ to either $nS_{1/2}$ or $nD_{5/2,3/2}$ Rydberg states. The coupling strength can be expressed in terms of the Rabi frequency $\Omega = -\mu \cdot E/\hbar$, which scales linearly with the dipole matrix element. For experiments where the coupling Rabi frequency is to be kept constant over a range of n, it is necessary to calculate the dipole matrix elements for the transition. Using the core potential and the energy of the $5P_{3/2}$ state,[1] an approximate $5P_{3/2}$ wavefunction can be calculated to find the radial dipole matrix elements $\langle 5P_{3/2}|er|n\ell j \rangle$ for the allowed transitions. The results are plotted in Fig. 2.2, showing a stronger coupling to the $nD_{5/2}$ state. The matrix elements are around five orders of magnitude weaker than the coupling to the nearest Rydberg states ($\sim 1000\,ea_0$ at $n = 40$), and are fitted using the scaling $C_\ell n^{\star-3/2}$ to obtain the coefficients $C_S = 4.502\,ea_0$ and $C_D = 8.457\,ea_0$, in good agreement with Deiglmayr et al. [18].

The total matrix element is obtained by multiplying the radial part by the angular component. For transition between the stretched states with $j = \ell + 1/2$, $|m_j| = j$, the angular coupling of Eq. 2.16 reduces to

---

[1] Below $n \sim 20$ the quantum defects give poor agreement as the electron has a strong interaction with the core.

$$\langle P_{3/2}, m_j = 3/2|\mu_q|\ell' j' m'_j\rangle = \sqrt{\frac{\ell_{max}}{(2\ell_{max} + 1)}}, \tag{2.17}$$

giving $\sqrt{1/3}$ for transitions to $nS_{1/2}, m_j = 1/2$ and $\sqrt{2/5}$ to $nD_{5/2}, m_j = 5/2$, further enhancing the coupling to $nD_{5/2}$ relative to $nS_{1/2}$.

2.4 Stark Shift

Applying a static electric field E along the z-axis causes the states to mix, shifting the energy levels relative to the bare atom, known as the Stark shift. To calculate the atomic energy states in the presence of an electric field, it is necessary to find the eigenvalues of the Stark Hamiltonian [14]

$$\mathcal{H}_{Stark} = \mathcal{H}_{atom} + E\hat{z}. \tag{2.18}$$

The electric field term $E\hat{z}$ creates off-diagonal couplings between states, with the selection rule $\Delta m_j = 0$ such that $|m_j|$ states are coupled together. The new energy levels are found by diagonalising $\mathcal{H}_{Stark}$ as a function of E for all states with a given $|m_j|$ to create an energy diagram known as a Stark map.

Figure 2.3 shows Stark maps calculated at $n = 40$ for the $|m_j| = 1/2$ and $5/2$ manifolds. The angular momentum states are truncated at $\ell = 20$ as this is sufficient for convergence of the energy levels of the states for $\ell \leq 3$. From (a), the effect of the quantum defects in shifting the energy levels is clear, as the closest $S_{1/2}$ state to the $n = 40$ hydrogenic manifold is $43S_{1/2}$. The high-ℓ hydrogenic states are degenerate, leading to a first-order linear Stark shift. In the $|m_j| = 1/2$ states, all of the levels are coupled leading to avoided crossings between the states with closest ℓ. In (b), the $|m_j| = 5/2$ hydrogenic states separate into $|m_\ell| = 2, 3$ states. This is the relevant quantum number as the electric field couples to ℓ, leading to a mixture of real and avoided crossings observable between adjacent n states.

2.4.1 Scalar Polarisability

At low fields, the Stark effect acts as a second-order perturbation on the states with $\ell \leq 3$ to give a quadratic shift of the form

$$\Delta W = -\frac{1}{2}\alpha_0 E^2, \tag{2.19}$$

where α_0 is the static polarisability, which for state $|n, \ell, j, m_j\rangle$ is given by

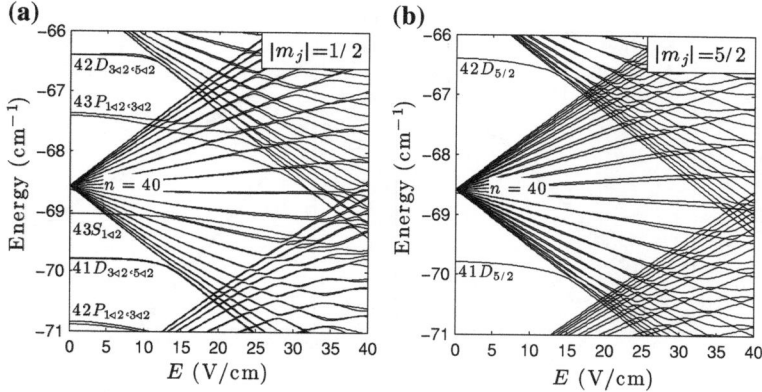

Fig. 2.3 $n = 40$ Stark maps for Rb. **a** $|m_j| = 1/2$ manifold shows avoided crossings between states with $\Delta \ell = \pm 1$. **b** $|m_j| = 5/2$. Hydrogenic states are split into $|m_\ell| = 2, 3$ manifolds, resulting in a mixture of avoided and real crossings between adjacent n states.

Table 2.2 Parameters for calculating static polarisability $\alpha_0 = \beta_1 n^{*6} + \beta_2 n^{*7}$ in units of MHz/(V/cm)^2

State	$|m_j|$	β_1 ($\times 10^{-9}$)	β_2 ($\times 10^{-11}$)	State	$|m_j|$	β_1 ($\times 10^{-9}$)	β_2 ($\times 10^{-8}$)
$S_{1/2}$	1/2	2.188	5.486	$F_{1/2}$	1/2	−1.655	1.612
$P_{1/2}$	1/2	2.039	51.456	$F_{1/2}$	3/2	−1.308	1.350
$P_{3/2}$	1/2	2.449	62.011	$F_{1/2}$	5/2	−0.634	0.826
$P_{3/2}$	3/2	1.611	52.948	$F_{1/2}$	1/2	−1.624	1.623
$D_{3/2}$	1/2	2.694	−6.159	$F_{1/2}$	3/2	−1.457	1.478
$D_{3/2}$	3/2	1.725	22.259	$F_{1/2}$	5/2	−1.077	1.188
$D_{5/2}$	1/2	2.770	−12.223	$F_{1/2}$	7/2	−0.530	0.753
$D_{5/2}$	3/2	2.352	1.772				
$D_{5/2}$	5/2	1.513	29.763				

$$\alpha_0 = \sum_{n',\ell',j' \neq n,\ell,j} \frac{|\langle n, \ell, j, m_j|\mu_0|n', \ell', j', m_j\rangle|^2}{W_{n'\ell'j'} - W_{n\ell j}}. \tag{2.20}$$

The polarisability $\alpha_0 \simeq \mu^2/\Delta$, where $\Delta \propto n^{*-3}$ is the energy of the nearest state and $\mu \propto n^{*2}$, giving $\alpha_0 \propto n^{*7}$. Consequently Rydberg states are incredibly sensitive to electric fields, allowing precise control over the Rydberg energy levels and making them suitable for applications in electrometry [19–21].

The static polarisabilities can be obtained experimentally by fitting the low-field dependence of the energy-levels for each state. To test the accuracy of the code, the polarisabilities calculated from Eq. 2.20 for the $nS_{1/2}$ states are compared to the measurements of O'Sulivan et al. [22]. The results are plotted in Fig. 2.4a, showing excellent agreement between theory and experiment. In [22] the authors fit the data to an empirical scaling of the form

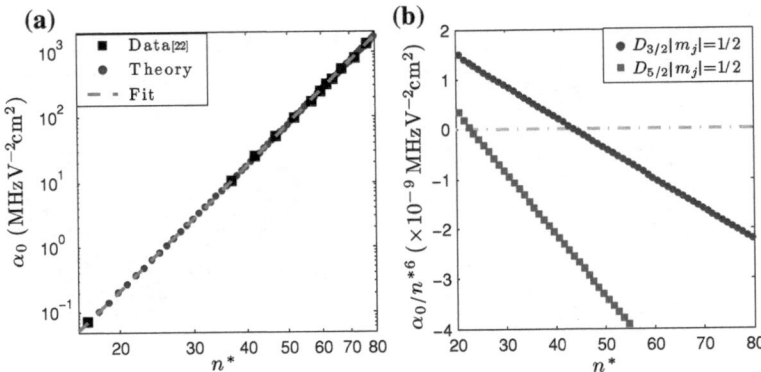

Fig. 2.4 Scalar polarisability. **a** Comparison of calculated $n S_{1/2}$ static polarisabilities α_0 to experimental data from Ref. [22]. **b** The static polarisability for the $D_{5/2,3/2}$ $|m_j| = 1/2$ states changes sign, resulting in a blue-shift at low fields

$$\alpha_0 = \beta_1 n^{*6} + \beta_2 n^{*7}, \qquad (2.21)$$

where α_0 is in units of MHz/(V/cm)^2, obtaining $\beta_1 = 2.202 \times 10^{-9}$ and $\beta_2 = 5.53 \times 10^{-11}$ for the measured data. Table 2.2 shows the results obtained from least-square fitting this scaling to the calculated polarisabilities over the range $n = 20-100$ for all states with $\ell \le 3$, which are consistent these empirical values for the $n S_{1/2}$ states. For $|m_j| = 1/2$ in the D states, the static polarisability is initially positive at low n and changes sign to become negative for the higher excited states, shown in Fig. 2.4b. This gives a positive Stark shift at low field for states above $24 D_{5/2}$. However, as the electric field increases, the D-states have an avoided crossing with the F-states and the energy shift becomes negative again. This can be seen from the Stark map in Fig. 2.3a.

2.5 Summary

The Rydberg series describes a set of states with simple scaling laws for fundamental properties such as transition frequencies, radiative lifetime or static polarisability in terms of the principal quantum number, which can be derived from the analytic solutions for the wavefunctions of hydrogen. For the alkali metal atoms, the interaction with the core creates a perturbation to the hydrogenic states that is characterised by the quantum defects. Using a model potential, the wavefunctions can be obtained numerically, enabling calculation of the transition dipole matrix elements between the states. From these matrix elements a wide range of properties can be calculated, such as the electric field sensitivity as described above. The most important property of the Rydberg states is the large dipole moment for transitions to adjacent Rydberg states $\propto n^{*2}$. As will be shown in the following chapter, this leads to very strong interactions between a pair of atoms excited to the Rydberg state.

References

1. J.R. Rydberg, On the structure of the line-spectra of the chemical elements. Phil. Mag. **29**(179), 331 (1890)
2. N. Bohr, On the constitution of atoms and molecules. Phil. Mag. **26**, 1 (1913)
3. E. Schrödinger, An undulatory theory of the mechanics of atoms and molecules. Phys. Rev. **28**(6), 1049 (1926)
4. M. Saffman, T.G. Walker, K. Mølmer, Quantum information with Rydberg atoms. Rev. Mod. Phys. **82**(3), 2313 (2010)
5. T.F. Gallagher, Rydberg atoms. Rep. Prog. Phys. **51**(2), 143 (1988)
6. T.F. Gallagher, *Rydberg Atoms* (CUP, New York, 2005)
7. W. Li, I. Mourachko, M.W. Noel, T.F. Gallagher, Millimeter-wave spectroscopy of cold Rb Rydberg atoms in a magneto-optical trap: quantum defects of the ns, np and nd series. Phys. Rev. A **67**, 052502 (2003)
8. J. Han, Y. Jamil, D.V.L. Norum, P.J. Tanner, T.F. Gallagher, Rb nf quantum defects from millimeter-wave spectroscopy of cold ^{85}Rb Rydberg atoms. Phys. Rev. A **74**(5), 054502 (2006)
9. M. Marinescu, H.R. Sadeghpour, A. Dalgarno, Dispersion coefficients for alkali-metal dimers. Phys. Rev. A **49**(2), 982 (1994)
10. C.E. Theodosiou, Lifetimes of alkali-metal-atom Rydberg states. Phys. Rev. A **30**(6), 2881–2909 (1984)
11. S.A. Bhatti, C.L. Cromer, W.E. Cooke, Analysis of the Rydberg character of the $5d7d\,D21$ state of barium. Phys. Rev. A **24**(1), 161–165 (1981)
12. B.V. Numerov, A method of extrapolation of perturbations. MNRAS **84**, 592 (1924)
13. B.V. Numerov, Note on the numerical integration of $d^2x/dt^2 = f(x, t)$. Astron. Nachr. **230**, 359 (1927)
14. M.L. Zimmerman, M.G. Littman, M.M. Kash, D. Kleppner, Stark structure of the Rydberg states of alkali-metal atoms. Phys. Rev. A **20**(6), 2251 (1979)
15. B.H. Bransden, C.J. Joachain, *Physics of Atoms and Molecules* (Longman Scientific and Technical, London, 1983)
16. M. Weissbluth, *Atoms and Molecules*, (Academic Press, New York, 1978)
17. I.I. Sobelman, *Atomic Spectra and Radiative Transitions* (Springer, Berlin, 1979)
18. J. Dieglmayr, M. Reetz-Lamour, T. Amthor, S. Westermann, A.L. de Oliveira, M. Weidemüller, Coherent excitation of Rydberg atoms in an ultracold gas. Opt. Comm. **264**, 293 (2006)
19. J. Neukammer, H. Rinneberg, K. Vietzke, A. König, H. Hieronymus, M. Kohl, H.J. Grabka, G. Wunner, Spectroscopy of Rydberg atoms at $n \approx 500$: observation of Quasi–Landau resonances in low magnetic fields. Phys. Rev. Lett. **59**(26), 2947 (1987)
20. A. Osterwalder, F. Merkt, Using high Rydberg states as electric field sensors. Phys. Rev. Lett. **82**(9), 1831 (1999)
21. A. Tauschinsky, R.M.T. Thijssen, S. Whitlock, H.B. van Linden, van den Heuvell, R.J.C. Spreeuw, Spatially resolved excitation of Rydberg atoms and surface effects on an atom chip. Phys. Rev. A **81**, 063411 (2010)
22. M.S. O'Sullivan, B.P. Stoicheff, Scalar polarizabilities and avoided crossings of high Rydberg states in Rb. Phys. Rev. A **31**(4), 2718 (1985)

Chapter 3
Rydberg Atom Interactions

As discussed in the introduction, the strong dipole–dipole interactions of the Rydberg states make them ideal for studying quantum many body physics, and applications in quantum information [1]. One of the main advantages of Rydberg atoms over other dipolar systems, such as polar molecules [2, 3], is the ability to control the strength, sign and spatial dependence through choice of state, in addition to be being able to turn the interactions off by returning population to the ground state. This chapter outlines the principle behind dipole interactions of the Rydberg states, detailing the properties of the $S_{1/2}$ and $D_{5/2}$ states of rubidium.

3.1 Dipole–dipole Interactions

Consider a pair of atoms initially in state $|r\rangle = |n, \ell, j, m_j\rangle$ separated by distance $\boldsymbol{R}$, shown schematically in Fig. 3.1a. The dipole–dipole interaction energy for this system can be written in atomic units as

$$V(\boldsymbol{R}) = \frac{\boldsymbol{\mu}_1 \cdot \boldsymbol{\mu}_2}{R^3} - \frac{3(\boldsymbol{\mu}_1 \cdot \boldsymbol{R})(\boldsymbol{\mu}_2 \cdot \boldsymbol{R})}{R^5}, \tag{3.1}$$

where $\boldsymbol{\mu}_{1,2}$ are the dipole moments associated for the transitions from $|r\rangle$ to $|r'\rangle$ and $|r''\rangle$ respectively. Taking $\boldsymbol{R}$ along the z-axis ($\theta = 0$), the dipole–dipole interaction reduces to

$$V(R) = \frac{\mu_{1-}\mu_{2+} + \mu_{1+}\mu_{2-} - 2\mu_{1z}\mu_{2z}}{R^3}, \tag{3.2}$$

where μ_{iq} denotes the dipole operator of atom $i = \{1, 2\}$ and subscript $q = \{-, z, +\}$ corresponds to a $\{\sigma^+, \pi, \sigma^-\}$ transition. In this geometry the selection rules for the dipole–dipole interactions preserve the total angular momentum $M = m_{j1} + m_{j2}$ of the initial pair states.

J. D. Pritchard, *Cooperative Optical Non-Linearity in a Blockaded Rydberg Ensemble*, Springer Theses, DOI: 10.1007/978-3-642-29712-0_3, © Springer-Verlag Berlin Heidelberg 2012

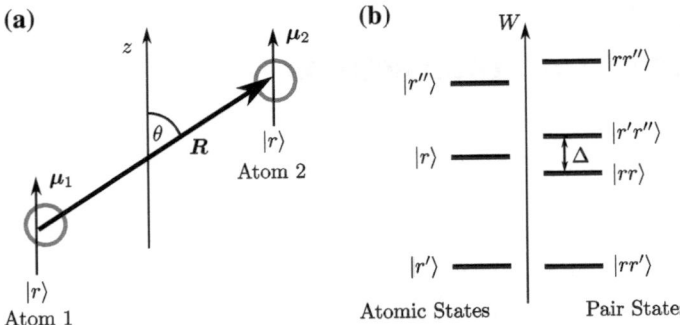

Fig. 3.1 **a** Dipole–dipole interactions between two atoms with interatomic separation R at an angle θ to the z-axis. **b** Transformation from atomic to pair state basis reveals near-resonant states with an energy defect Δ that are coupled by the dipole–dipole interaction

To calculate the energy shift due to the dipole–dipole interaction, it is necessary to transform from an atomic basis to a pair basis, as illustrated in Fig. 3.1b. The initial pair state $|rr\rangle$ is coupled by $V(R)$ to a state $|r'r''\rangle$ which has an energy defect Δ given by

$$\Delta = W_{|r'\rangle} + W_{|r''\rangle} - 2W_{|r\rangle},$$ (3.3)

which represents the energy difference of the pair states at infinite separation. The Hamiltonian describing the dipole–dipole interaction for the basis states $|rr\rangle$, $|r'r''\rangle$ is

$$\mathcal{H} = \begin{pmatrix} 0 & V(R) \\ V(R) & \Delta \end{pmatrix}.$$ (3.4)

The eigenvalues of this Hamiltonian are

$$\lambda_{\pm} = \frac{\Delta \pm \sqrt{\Delta^2 + 4V(R)^2}}{2},$$ (3.5)

such that the energy of the pair states is now dependent upon the separation of the two atoms.

The form of the spatial dependence can be derived in two distinct limits:

(i) **Long range** $(V(R) \ll \Delta)$

$$\Delta W = -\frac{V(R)^2}{\Delta} = -\frac{C_6}{R^6}.$$ (3.6)

This is the van der Waals (vdW) regime where the sign of the interaction is determined by Δ. In this limit, the strength of the interaction is characterized by parameter C_6 which scales proportional to n^{*11} as $V(R) \propto \mu^2 \propto n^{*4}$ and the energy defect $\Delta \propto n^{*-3}$.

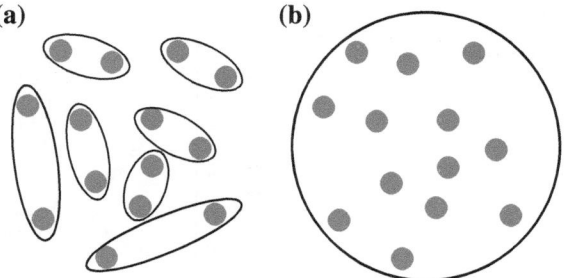

Fig. 3.2 Many-body interactions. **a** For the case of many-body interactions, in the van der Waals regime $V \propto 1/R^6$ interactions are dominated by nearest-neighbour allowing consideration of isotropic pair-wise interactions. **b** For the resonant dipole regime $V \propto 1/R^3$ however, all of the surrounding atoms are equally important and it is now necessary to consider the full many-body system

(ii) **Short range** $(V(R) \gg \Delta)$

$$\Delta W = \pm V(R) = \pm \frac{C_3}{R^3}. \tag{3.7}$$

This is the resonant dipole–dipole regime as it has a $1/R^3$ dependence associated with a pair of static dipoles, scaling as $C_3 \propto n^{*4}$.

The transition between the $1/R^3$ and $1/R^6$ regimes occurs at the van der Waals radius when $V(R_{\text{vdW}}) = \Delta$, where $R_{\text{vdW}} = \sqrt[6]{|C6/\Delta|} \propto n^{*7/3}$.

An important difference between these two regimes is the contribution of the nearest-neighbour to the total interaction when considering a many-body atomic ensemble, as shown in Fig. 3.2. For a uniform density ρ, the average interatomic spacing is given by $R_{\text{avg}} = (5/9)\rho^{-1/3}$ [4]. Assuming the nearest neighbour is at this radius, the pair-wise interaction with this atom is $V_{\text{pair}} = V(R_{\text{avg}})$, whilst the interaction energy contributed by all other atoms in the system can be evaluated from,

$$V_s = \int_{R_{\text{avg}}}^{\infty} 4\pi R^2 \rho V(R) \, \mathrm{d}R. \tag{3.8}$$

In the van der Waals regime, this reduces to $V_s \simeq 0.7 V_{\text{pair}}$, showing interactions are dominated by the nearest neighbour, allowing the many-body system to be treated as an ensemble of interacting pairs. For the resonant dipole interaction however, the integral diverges as the $1/R^3$ interaction cancels with the R^3 scaling of the number of atoms at radius R. This means the contribution from the surrounding atoms is significantly larger than the closest atom, and the effect of all atoms must be included. A thorough discussion of this relative contribution of the surrounding atoms is given in [5, 6].

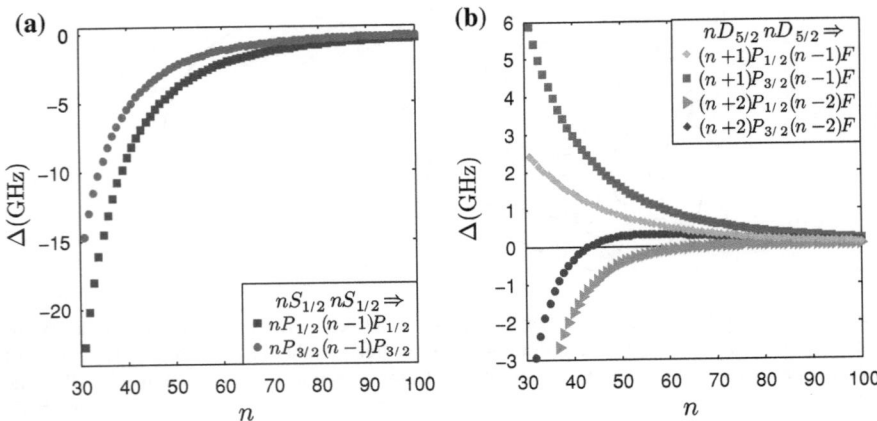

Fig. 3.3 Rubidium pair-state energy defects in zero field. **a** $nS_{1/2}nS_{1/2}$ pair states have $\Delta < 0$ for all $n'Pn''P$ states, leading to repulsive interactions. **b** $nD_{5/2}nD_{5/2}$ pair states have the smallest energy defect for the $(n+2)P(n-2)F$ states, which gives attractive long-range interactions for $n > 43$

3.2 Interaction Strengths

In the long range van der Waals regime, the interaction strength is dominated by the pair state with the smallest energy defect, which determines the sign of the interaction. However, to calculate the magnitude of the interaction it is necessary to consider not only the near-resonant pair state, but all combinations of states which are dipole coupled to the initial pair state. Figure 3.3 shows the energy defects as a function of n for (a) $nS_{1/2}nS_{1/2}$ and (b) $nD_{5/2}nD_{5/2}$ states in Rb. For the $S_{1/2}$ states, the smallest energy defect is given by the coupling to the $nP_{3/2}(n-1)P_{3/2}$, which is negative for all n. Thus $C_6 < 0$, corresponding to repulsive interactions for the $S_{1/2}$ states.

In the $D_{5/2}$ states, there are a range of near-resonant dipole-coupled channels, all with energy defects much smaller than the equivalent S states. The dominant coupling is to $(n+2)P_{3/2}(n-2)F$, which changes sign at $n > 43$, changing from repulsive to attractive interactions. In addition to smaller Δ, the matrix elements of the $D_{5/2} \rightarrow P, F$ are larger than the $S_{1/2} \rightarrow P$, leading to stronger interactions with a larger R_{vdW} [7]. For example, the $60D_{5/2}60D_{5/2}$ $M = 5$ pair state has $C_6 = 210\,\text{GHz}\,\mu\text{m}^6$, compared to $-140\,\text{GHz}\,\mu\text{m}^6$ for the $60S_{1/2}60S_{1/2}M = 1$ pair state.

Figure 3.4a shows the calculated pair potential for the $60S_{1/2}60S_{1/2}$ state obtained by diagonalisation of Eq. 3.4 for all pair states with $|\Delta| < 25\,\text{GHz}$, showing the effective splitting of the near-resonant $60P_{3/2}59P_{3/2}$ state. The coefficients C_3 and C_6 are determined by fitting the potential curve either side of $R_{vdW} = 2.1\,\mu\text{m}$. This gives $C_3 = -14.3\,\text{GHz}\,\mu\text{m}^3$ and $C_6 = -140\,\text{GHz}\,\mu\text{m}^6$, in excellent agreement with [8] which calculates C_6 using second-order perturbation theory in the uncoupled basis. In (b) the interaction potentials for a range of n are plotted, illustrating both the

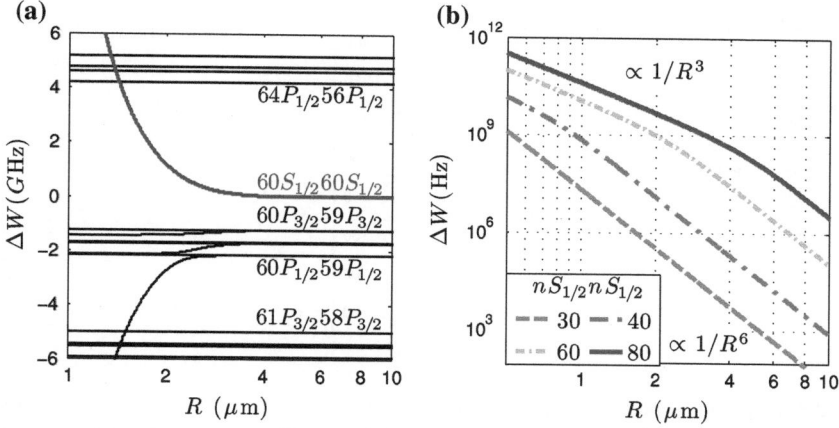

Fig. 3.4 $nS_{1/2}nS_{1/2}$ interaction potentials **a** $60S_{1/2}60S_{1/2}$ pair state energy calculated including all pair states with $\Delta < 25$ GHz. The interaction is dominated by the $60P_{3/2}59P_{3/2}$ state which is repelled by the coupling. **b** Comparison of interaction curves as a function of n, showing shift of the van-der-Waals radius R_{vdW} to larger R as n increases, with the $1/R^3$ resonant dipole interaction becoming relevant at densities of 10^{10} cm^{-3} for $n \gtrsim 80$

increased interaction strength with n and also the transition from $1/R^6$ to $1/R^3$ at short range. For a typical experiment density of 10^{10} cm^{-3}, the average pair separation $R_{avg} = 2.5$ μm. This corresponds to interactions in the van der Waals regime for $n \lesssim 60$, which is the largest n state used in work presented in this thesis. Therefore all interactions can be modelled as $V(R) = C_6/R^6$ in later sections.

3.3 Angular Dependence

In the discussion above it was assumed the dipoles were aligned with $\theta = 0$. More generally, the dipole–dipole coupling is a function of both θ and R given by [9]

$$
\begin{aligned}
V(R, \theta) = {} & \frac{\mu_{1-}\mu_{2+} + \mu_{1+}\mu_{2-} + (1 - 3\cos^2\theta)\mu_{1z}\mu_{2z}}{R^3} \\
& + \frac{3/2\sin^2\theta(\mu_{1+}\mu_{2+} + \mu_{1+}\mu_{2-} + \mu_{1-}\mu_{2+} + \mu_{1-}\mu_{2-})}{R^3} \\
& + \frac{3/\sqrt{2}\sin\theta\cos\theta(\mu_{1z}\mu_{2+} + \mu_{1z}\mu_{2-} + \mu_{1+}\mu_{2z} + \mu_{1-}\mu_{2z})}{R^3},
\end{aligned}
\tag{3.9}
$$

where for $\theta \neq 0$ the total angular momentum M of the initial pair state is no longer conserved, allowing states of different M to be coupled together. Figure 3.5 shows the van der Waals interaction strength for pairs of atoms in the (a) $60S_{1/2}60S_{1/2}$ and (b) $60D_{5/2}60D_{5/2}$ states as a function of θ. From (a) it is obvious the $S_{1/2}$ state

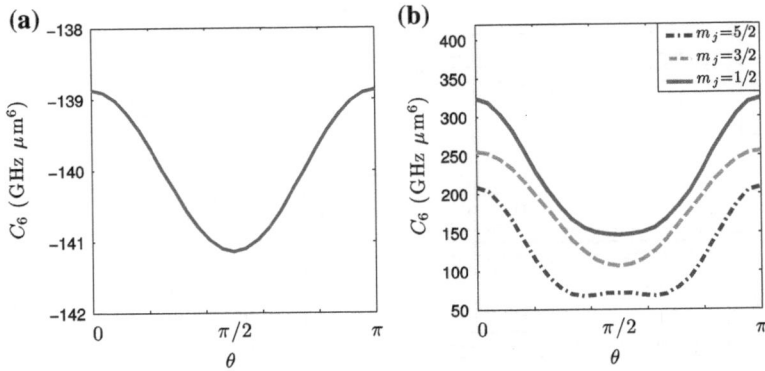

Fig. 3.5 Angular dependence of C_6 on θ for **a** $60S_{1/2}60S_{1/2}$ and **b** $60D_{5/2}60D_{5/2}$ for pairs of atoms in state m_j. This shows the $S_{1/2}$-states are almost isotropic whilst $D_{5/2}$-states have >50% variation with θ. The interaction is larger for $m_j = 1/2$ as it allows coupling to the $(n+2)P_{1/2}(n-2)F_{5/2}$ channel which has a smaller energy defect at $n=60$ (see Fig. 3.3b)

interactions are almost perfectly isotropic. This occurs because the dipole–dipole interaction couples to the orbital angular momentum, ℓ, which is zero for the S state pair and has a spherically symmetric distribution. The slight perturbation occurs from the fine-structure splitting of the P states. For the $D_{5/2}$ states however, the orbital angular momentum of the initial pair state is four, giving a significantly anisotropic interaction which is reduced by more than 50% at $\theta = \pi/2$ from the value aligned along z ($\theta = 0$).

3.4 Tuning Interactions with External Fields

The strength and sign of the dipole–dipole interaction can be controlled by choice of the $n\ell$ state of the initial Rydberg pair states, however there will always be a transition from the $1/R^3$ to $1/R^6$ regime. Choosing a very high principal quantum number can extend the transition radius R_{vdW} to large R, however at the cost of a weaker coupling to the intermediate state which scales as $\propto n^{*-3/2}$ (Fig. 2.2). An alternative approach is to tune the interaction by applying an external field to the system.

3.4.1 Static Electric Field

A static electric field causes a Stark shift of the Rydberg states as shown in Sect. 2.4 which can be used to tune the sign and magnitude of the energy defect, for example from attractive to repulsive interactions. A Förster resonance occurs when $\Delta = 0$,

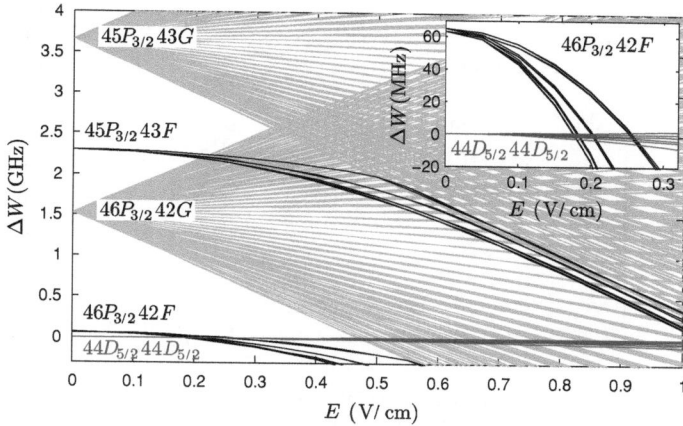

Fig. 3.6 Energy defect for states coupled to the $44D_{5/2}44D_{5/2}$ pair state as a function of static field, revealing Förster resonances around 0.2 V/cm for the near-resonant $46P_{3/2}42F$ states and around 1.1 V/cm for $45P_{3/2}43F$ pair states. This figure shows interactions for all possible M states, with the hydrogenic manifold plotted to illustrated the states causing the shift

leading to $1/R^3$ resonant dipole–dipole interactions for all R [10, 11]. An example is shown in Fig. 3.6 for the $44D_{5/2}44D_{5/2}$ pair state, where two Förster resonances are visible, occurring at 0.2 V/cm for the $46P_{3/2}42F$ state and at 1.1 V/cm for the $45P_{3/2}43F$ state. The shape of the pair potentials is dominated by the linear shift of the hydrogen-like states with $\ell > 3$ which are plotted to illustrate the origin of the linear shift of F pair states at high field.

Calculation of the interaction potentials in an applied field reveals the presence of long-range ($\sim$9 μm) molecular bound states known as macrodimers [12], which have been observed experimentally in Cs [13]. If the applied field is increased further, the Rydberg atoms will have a permanent rather than induced dipole moment, where the permanent dipole moment is given by the gradient $\mu = -\,\mathrm{d}W(E)/\,\mathrm{d}E$.

3.4.2 Microwave Dressing

The energy separation between close-lying Rydberg states typically corresponds to microwave frequencies, *e.g.* $46S_{1/2} \rightarrow 45P_{3/2} \sim 40\,\text{GHz}$. Consider a microwave field with detuning $\Delta_\mu = \omega_\mu - \omega_0$ tuned close to resonance with the transition from states $|r\rangle$ to $|r'\rangle$, with Rabi frequency Ω_μ (see Sect. 4.1). As the microwave field couples the single-atom states, whilst the dipole–dipole interaction couples the pair states, all four pair states must be considered, as shown in Fig. 3.7a. The Hamiltonian for the interacting system in the basis $\{|rr\rangle, |rr'\rangle, |r'r\rangle, |r'r'\rangle\}$ is

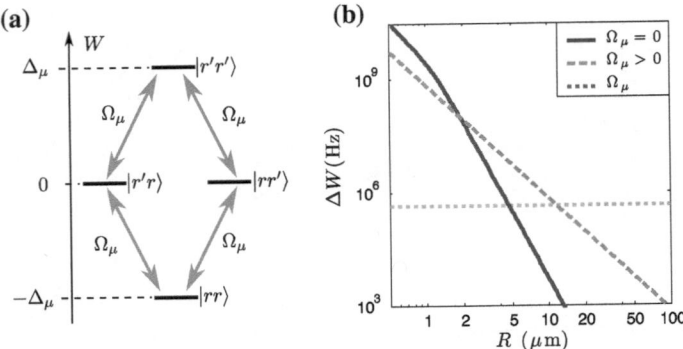

Fig. 3.7 Microwave-dressing of dipole–dipole interaction. **a** Applying a near-resonant microwave coupling between $|r\rangle$ and $|r'\rangle$ dresses the pair states giving a new pair basis separated by $\Delta_\mu \ll \Delta$. **b** Interaction potential for $46S_{1/2}46S_{1/2}$ pair-state with a resonant ($\Delta_\mu = 0$) microwave coupling of $\Omega_\mu/2\pi = 500$ kHz to the $45P_{1/2}$ state. The result is a $1/R^3$ behaviour for all R (*green line*), contrasting with the case for no microwave field (*blue line*). *Dashed line* shows energy of microwave splitting

$$\mathscr{H}_{\text{dd}} = \hbar \begin{pmatrix} -\Delta_\mu & \Omega_\mu/2 & \Omega_\mu/2 & V(R) \\ \Omega_\mu/2 & 0 & V(R) & \Omega_\mu/2 \\ \Omega_\mu/2 & V(R) & 0 & \Omega_\mu/2 \\ V(R) & \Omega_\mu/2 & \Omega_\mu/2 & \Delta_\mu \end{pmatrix}, \tag{3.10}$$

where the system has be transformed into the frame rotating at ω_μ. In this frame, the energy defect Δ has been replaced by the microwave detuning Δ_μ. Therefore the microwave field now chooses which Rydberg state ($|r'\rangle$) should contribute to the the dipole–dipole interactions. On resonance ($\Delta_\mu = 0$) the eigenvalues of $\mathscr{H}_{\text{dd}}$ are $\lambda = -V(R)$, $V(R) \pm \Omega_\mu$, corresponding to an Autler–Townes [14] splitting from the microwave dressing. The interaction is a resonant dipole–dipole $\propto 1/R^3$ for all R, shown in Fig. 3.7b for the $46S_{1/2}$ state, calculated for a resonant coupling to the $45P_{1/2}$ state with a dipole matrix element of $\mu_{s \to p} = \sqrt{2/9} \times 1924\, ea_0$ and $\Omega_\mu/2\pi = 500$ kHz. Plotted on the same graph is the interaction curve for the undressed van der Waals interactions. This clearly shows that for $R > 2\,\mu$m the microwave dressing enhances the strength of the dipole interaction, although it only exceeds the microwave splitting of Ω_μ for $R < 15\,\mu$m.

The advantage of the microwave dressing over applying a weak electric field is two-fold. Firstly, any pair state can be tuned into resonance rather than just the close lying states. Secondly, stabilising the microwave frequency to <1 Hz is much easier than controlling weak electric fields due to the stray electric field of around 50 mV/cm typical for cold atom experiments [6].

3.5 Dynamic Effects of Dipole–dipole Interactions

For an ensemble of Rydberg atoms, the dipole–dipole interactions play an important role in determining the dynamics of the system. The focus in this work is the blockade effect that suppresses excitation (see Sect. 5.3), however it is important to consider other effects of interactions. The most important process is ionisation of the Rydberg states due to collisions, resulting in Penning ionisation of the form

$$Rb^* + Rb^* \rightarrow Rb^{**} + Rb^+ + e^- \tag{3.11}$$

where the remaining atom is transferred into a different Rydberg state with $n' \lesssim n/\sqrt{2}$ from energy conservation [15]. For D-states, the attractive interactions cause atoms to be accelerated towards each other, leading to very rapid ionisation on timescales of a few μs [16, 17]. This rapid ionisation can trigger plasma formation as electrons become trapped in the attractive potential of the slow moving ions, leading to avalanche ionisation [18, 19]. To overcome this rapid ionisation, the repulsive $S_{1/2}$ states can be used. However ionisation still occurs due to collisions or from ℓ-changing due to absorption of a black-body photon, transferring the pair state onto an attractive potential [20], although at a lower rate than for the D state. Interestingly, the frequency of the excitation laser plays an important role in determining the ionisation dynamics—if the laser has the opposite detuning to the sign of the interaction, only long-range pairs are excited which take a long time to collide. For a laser detuning equal to the interaction shift, it is now the short-range pairs that are excited leading to rapid ionisation [17, 19]. Thus it is possible to control the initial separation of the Rydberg pair states in the system by choice of excitation frequency. This effect has been used to map out the nearest-neighbour distribution for a cold atom cloud [21].

3.6 Summary

The large matrix elements of the Rydberg states leads to strong dipole–dipole interactions between atoms. The interactions can be expressed in two regimes; at long range ($V(R) \ll \Delta$) this is the van der Waals regime with $\Delta W = -C_6/R^6$, whilst at short range ($V(R) \gg \Delta$) the atoms experience resonant dipole–dipole interactions $\Delta W = C_3/R^3$. For a typical experiment density, the average atomic separation $R_{avg} > R_{vdW}$, corresponding to van der Waals interactions for $n \lesssim 60$. Interactions play an important role in the dynamics of the system, with attractive interactions resulting in enhanced ionisation compared to the repulsive S-states. The scaling with n allows a great degree of control over the magnitude and sign of the interaction, which can be additionally tuned using external fields to create long range resonant dipole–dipole interactions. In Chap. 5, the cooperative nature of these interactions will be explored.

References

1. M. Saffman, T.G. Walker, K. Mølmer, Quantum information with Rydberg atoms. Rev. Mod. Phys. **82**(3), 2313 (2010)
2. D. DeMille, Quantum computation with trapped polar molecules. Phys. Rev. Lett. **88**(6), 067901 (2002)
3. R.M. Rajapakse, T. Bragdon, A.M. Rey, T. Calarco, S.F. Yelin, Single-photon nonlinearities using arrays of cold polar molecules. Phys. Rev. A **80**(1), 013810 (2009)
4. P. Hertz, Über den gegenseitigen durchschnittlichen Abstand von Punkten, die mit bekannter mittlerer Dichte im Raume angeordnet sind. Math. Ann. **67**, 387 (1909)
5. T. Amthor, J. Denskat, C. Giese, N.N. Bezuglov, A. Ekers, L. Cederbaum, M. Weidemuller, Autoionization of an ultracold Rydberg gas through resonant dipole coupling. Eur. Phys. J. D. **53**(3), 329 (2009)
6. T. Amthor. Interaction–Induced dynamics in ultracold Rydberg gases—mechanical effects and coherent processes. Ph.D. Thesis, Fakultät für Mathematik und Physik, Albert-Ludwigs-Universität Freiburg, 2008
7. T.G. Walker, M. Saffman, Consequences of Zeeman degeneracy for the van der Waals blockade between Rydberg atoms. Phys. Rev. A **77**(3), 032723 (2008)
8. K. Singer, J. Stanojevic, M. Weidemüller, R. Coté, Long-range interactions between alkali Rydberg atom pairs correlated to the ns–ns, np–np and nd–nd asymptotes. J. Phys. B **38**(2), S295 (2005)
9. A. Reinhard, T. Cubel Liebisch, B. Knuffman, G. Raithel, Level shifts of rubidium Rydberg states due to binary interactions. Phys. Rev. A **75**(3), 032712 (2007)
10. T.F. Gallagher, K.A. Safinya, F. Gounand, J.F. Delpech, W. Sandner, R. Kachru, Resonant Rydberg-atom—Rydberg-atom collisions. Phys. Rev. A **25**(4), 1905 (1982)
11. T. Vogt, M. Viteau, J. Zhao, A. Chotia, D. Comparat, P. Pillet, Dipole blockade at Förster resonances in high resolution laser excitation of Rydberg states of cesium atoms. Phys. Rev. Lett. **97**(8), 083003 (2006)
12. A. Schwettmann, K.R. Overstreet, J. Tallant, J.P. Shaffer, Analysis of long-range Cs Rydberg potential wells. J. Mod. Opt. **54**(17), 2551 (2007)
13. K.R. Overstreet, A. Schwettmann, J. Tallant, D. Booth, J.P. Shaffer, Observation of electric-field-induced Cs Rydberg atom macrodimers. Nature Phys. **5**, 581 (2009)
14. S.H. Autler, C.H. Townes, Stark effect in rapidly varying fields. Phys. Rev. **100**(2), 703 (1955)
15. T.F. Gallagher, P. Pillet, Dipole–dipole interactions of Rydberg atoms. Adv. At. Mol. Opt. Phys. **56**, 161 (2008)
16. W. Li, P.J. Tanner, T.F. Gallagher, Dipole–dipole excitation and ionization in an ultracold gas of Rydberg atoms. Phys. Rev. Lett. **94**(17), 173001 (2005)
17. T. Amthor, M. Reetz-Lamour, S. Westermann, J. Denskat, M. Weidemüller, Mechanical effect of van der Waals interactions observed in real time in an ultracold Rydberg gas. Phys. Rev. Lett. **98**(2), 023004 (2007)
18. M.P. Robinson, B. Tolra, B. Laburthe, M.W. Noel, T.F. Gallagher, P. Pillet, Spontaneous evolution of Rydberg atoms into an ultracold plasma. Phys. Rev. Lett. **85**(21), 4466 (2000)
19. W. Li, P.J. Tanner, Y. Jamil, T.F. Gallagher, Ionization and plasma formation in high n cold Rydberg samples. Eur. Phys. J. D. **40**(1), 27 (2006)
20. T. Amthor, M. Reetz-Lamour, C. Giese, M. Weidemüller, Modeling many-particle mechanical effects of an interacting Rydberg gas. Phys. Rev. A **76**(5), 054702 (2007)
21. T. Amthor, C. Giese, C.S. Hofmann, M. Weidemüller, Evidence of antiblockade in an ultracold Rydberg gas. Phys. Rev. Lett. **104**, 013001 (2010)

Chapter 4
Atom-Light Interactions

The simplest case in which to consider the interaction between atoms and light is that of a two-level atom driven by a coherent optical field. This system has been exhaustively studied e.g. [1, 2], revealing a range of coherent effects such as Rabi oscillations [3] and trapping due to the optical dipole force [4, 5]. Typically, the excited state in the two-level system has a finite lifetime due to spontaneous emission back to the ground state. On one hand this decay is advantageous, as it allows atoms to be cooled by radiation pressure [6–8]. On the other hand, the susceptibility is therefore dominated by a large, absorptive $\chi^{(1)}$ component [9]. The driven two-level system is thus poorly suited to applications in non-linear optics at the single-photon level.

However, the addition of a third level and a second optical field gives rise to a range of coherent phenomena including electromagnetically induced transparency (EIT) [10, 11] which suppresses the resonant absorption. The result is a very large dispersive optical non-linearity which can be used to control the propagation of light through the medium.

4.1 Three Level Atom

Consider a three-level atom with ground $|g\rangle$, excited $|e\rangle$ and Rydberg $|r\rangle$ states separated by energy $\hbar\omega_{eg}$ and $\hbar\omega_{re}$ respectively, as shown in Fig. 4.1 for a cascade, or ladder, configuration. The atom is driven by two laser fields; a probe laser field at frequency ω_p which drives the transition from $|g\rangle$ to $|e\rangle$ with detuning $\Delta_p = \omega_p - \omega_{eg}$, and a coupling laser field at frequency ω_c detuned by $\Delta_c = \omega_c - \omega_{re}$ from the $|e\rangle$ to $|r\rangle$ transition. The lasers are assumed to be classical monochromatic electric fields $\boldsymbol{E}_{p,c}(t) = \boldsymbol{E}_{p,c}\cos(\omega_{p,c}t)$ which couple to the electric dipole moment of the atom $\boldsymbol{d}$

$$\boldsymbol{d} = \boldsymbol{d}_{eg}(\hat{\pi}^+ + \hat{\pi}^-) + \boldsymbol{d}_{re}(\hat{\Xi}^+ + \hat{\Xi}^-), \tag{4.1}$$

J. D. Pritchard, *Cooperative Optical Non-Linearity in a Blockaded Rydberg Ensemble*,
Springer Theses, DOI: 10.1007/978-3-642-29712-0_4,
© Springer-Verlag Berlin Heidelberg 2012

Fig. 4.1 Three-level cascade
system with ground $|g\rangle$,
excited $|e\rangle$ and Rydberg
$|r\rangle$ states. A probe laser
at frequency ω_p drives the
$|g\rangle \rightarrow |e\rangle$ transition whilst a
coupling laser at frequency ω_c
couples levels $|e\rangle$ and $|r\rangle$

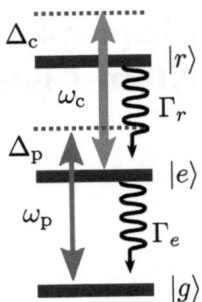

where $\mathbf{d}_{ij} = \langle i| - e\mathbf{r}|j\rangle$ is the dipole matrix element for the transition from $|i\rangle$ to $|j\rangle$
and the dipole operators $\hat{\pi}^{\pm}$, $\hat{\Xi}^{\pm}$ are the raising and lowering operators of the atomic
dipole for the two transitions, defined as

$$\hat{\pi}^+ = |e\rangle\langle g|, \ \hat{\pi}^- = |g\rangle\langle e|,$$
$$\hat{\Xi}^+ = |r\rangle\langle e|, \ \hat{\Xi}^- = |e\rangle\langle r|. \tag{4.2}$$

Applying the dipole-approximation[1] the coupling between the electric field and
the atom is $V = -\mathbf{d}\cdot(\mathbf{E}_p + \mathbf{E}_c)$, where the magnitude of the coupling can be expressed
in terms of the Rabi frequencies $\Omega_p = -\mathbf{E}_p \cdot \mathbf{d}_{eg}/\hbar$ and $\Omega_c = -\mathbf{E}_c \cdot \mathbf{d}_{er}/\hbar$ to give

$$V = \frac{\hbar\Omega_p}{2}(\hat{\pi}^- + \hat{\pi}^+) + \frac{\hbar\Omega_c}{2}(\hat{\Xi}^- + \hat{\Xi}^+), \tag{4.3}$$

where the rotating-wave approximation has been used to remove the non-resonant
terms corresponding to emission of a photon with an excitation of the atom and
absorption of a photon with de-excitation of the atom (see §A.11 of Ref. [2]). The
Hamiltonian for the coupled system is $\mathscr{H} = \mathscr{H}_A + V$, where $\mathscr{H}_A$ is the energy of
the bare atom

$$\mathscr{H} = -\hbar\Delta_p\hat{\pi}^+\hat{\pi}^- - \hbar(\Delta_p + \Delta_c)\hat{\Xi}^+\hat{\Xi}^-, \tag{4.4}$$

which acts on a wavefunction of the form $|\psi\rangle = a_g|g\rangle + a_e|e\rangle + a_r|r\rangle$. The states
$|g\rangle$, $|e\rangle$ and $|r\rangle$ can be expressed as orthogonal normalised column vectors

$$|g\rangle = \begin{pmatrix} 1 \\ 0 \\ 0 \end{pmatrix}, \ |e\rangle = \begin{pmatrix} 0 \\ 1 \\ 0 \end{pmatrix}, \ |r\rangle = \begin{pmatrix} 0 \\ 0 \\ 1 \end{pmatrix}, \tag{4.5}$$

from which the total Hamiltonian $\mathscr{H}$ is given in matrix form in this basis as

---

[1] Valid providing the electric field doesn't change rapidly over the length scale of the atom.

$$\mathscr{H} = \hbar \begin{pmatrix} 0 & \Omega_p/2 & 0 \\ \Omega_p/2 & -\Delta_p & \Omega_c/2 \\ 0 & \Omega_c/2 & -\Delta_p - \Delta_c \end{pmatrix}. \tag{4.6}$$

Using the Hamiltonian it is possible to calculate the dynamics in the absence of decoherence using the Schrödinger equation

$$i\hbar \frac{d}{dt}|\psi\rangle = \mathscr{H}|\psi\rangle. \tag{4.7}$$

For a real atomic system however, the excited states have a finite lifetime $\tau_{e,r}$ and it is necessary to treat the spontaneous emission of photons at rate $\Gamma_{e,r} = 1/\tau_{e,r}$. Spontaneous emission is a dissipative process and cannot be included in this Hamiltonian as a unitary process. Therefore the time evolution of the density matrix σ is used, instead of the wavefunction $|\psi\rangle$, to derive a master equation for the atom in which spontaneous decay can be included whilst preserving the normalisation.[2] The density operator for a pure state is defined as $\hat{\sigma} = |\psi\rangle\langle\psi|$, resulting in the density matrix given by

$$\sigma = \begin{pmatrix} |a_g|^2 & a_g a_e^* & a_g a_r^* \\ a_e a_g^* & |a_e|^2 & a_e a_r^* \\ a_r a_g^* & a_r a_e^* & |a_r|^2 \end{pmatrix} = \begin{pmatrix} \sigma_{gg} & \sigma_{ge} & \sigma_{gr} \\ \sigma_{eg} & \sigma_{ee} & \sigma_{er} \\ \sigma_{rg} & \sigma_{re} & \sigma_{rr} \end{pmatrix}. \tag{4.8}$$

To include the effect of spontaneous emission, the atom can be considered to couple to a reservoir initially in the vacuum state into which it can emit a photon, causing a relaxation of the atomic excitation. The coupling to the reservoir is described by the Lindblad superoperator $\mathcal{L}(\sigma)$ [13]

$$\mathcal{L}(\sigma) = -\frac{1}{2}\sum_m (C_m^\dagger C_m \sigma + \sigma C_m^\dagger C_m) + \sum_m C_m \sigma C_m^\dagger, \tag{4.9}$$

where the sum is over all decay modes m. For a given decay channel from $|i\rangle$ to $|j\rangle$, the first summation describes loss of population from state $|i\rangle$ due to emission of a photon, and the corresponding decay in the coherence terms $\sigma_{ji,ij}$, whilst the final term shows population being restored into state $|j\rangle$, ensuring $\mathrm{Tr}\{\sigma\} = 1$ for all times [14].

For the three-level atom there are two decay modes, one from $|e\rangle$ at rate Γ_e and another from $|r\rangle$ at rate Γ_r, which are described by the operators

$$C_e = \sqrt{\Gamma_e}|g\rangle\langle e|, \tag{4.10a}$$

$$C_r = \sqrt{\Gamma_r}|e\rangle\langle r|. \tag{4.10b}$$

---

[2] Alternatively a stochastic approach can be used to solve the Schrödinger equation for the wavefunction with dissipation [12].

Inserting these into Eq. 4.9, the Lindblad operator for the three-level atoms is

$$\mathcal{L}(\sigma) = \begin{pmatrix} \Gamma_e \sigma_{ee} & -\frac{1}{2}\Gamma_e \sigma_{ge} & -\frac{1}{2}\Gamma_r \sigma_{gr} \\ -\frac{1}{2}\Gamma_e \sigma_{eg} & -\Gamma_e \sigma_{ee} + \Gamma_r \sigma_{rr} & -\frac{1}{2}(\Gamma_e + \Gamma_r)\sigma_{er} \\ -\frac{1}{2}\Gamma_r \sigma_{rg} & -\frac{1}{2}(\Gamma_e + \Gamma_r)\sigma_{re} & -\Gamma_r \sigma_{rr} \end{pmatrix}. \tag{4.11}$$

The time evolution of the density matrix is calculated using the Liouville equation, which is the equivalent of the Schrödinger equation for the density matrix, where now the Lindblad operator can be included to account for spontaneous decay. The resulting equation, known as the master equation, or optical Bloch equation (OBE), is

$$\dot{\sigma} = \frac{i}{\hbar}[\sigma, \mathcal{H}] + \mathcal{L}(\sigma). \tag{4.12}$$

4.1.1 Finite Laser Linewidth

A nominally monochromatic source, such as a laser, does not emit at a single frequency, but instead has fluctuations in the emission frequency. Typically, the frequency spectrum of the fluctuations is assumed to be Lorentzian [15, 16]. The laser linewidth is therefore defined by the Lorentzian half-width at half maximum of the emission spectrum. The effect of this finite linewidth is to increase the dephasing rate of the off-diagonal coherence terms for the states coupled to the laser field, whilst leaving the diagonal populations unchanged.[3] Expressing the off-diagonal dephasing terms of the Lindblad operator in Eq. 4.11 as $\mathcal{L}(\sigma)_{ji} = -\gamma_{ji}\sigma_{ji}$, the effect of finite laser linewidth can be included by modifying the dephasing rates as follows [16, 17]

$$\gamma_{eg} \rightarrow \gamma_{eg} + \gamma_{\mathrm{p}}, \tag{4.13a}$$

$$\gamma_{rg} \rightarrow \gamma_{rg} + \gamma_{\mathrm{rel}}, \tag{4.13b}$$

$$\gamma_{re} \rightarrow \gamma_{re} + \gamma_{\mathrm{c}}, \tag{4.13c}$$

were $\gamma_{\mathrm{p,c}}$ are the linewidth of the probe and coupling lasers respectively and γ_{rel} is the linewidth of the two-photon resonance. For two independent lasers $\gamma_{\mathrm{rel}} = \gamma_{\mathrm{p}} + \gamma_{\mathrm{c}}$, which arises from the fact that the convolution of two Lorentzians of width $\gamma_{1,2}$ is equal to a Lorentzian whose width is $\gamma_1 + \gamma_2$. In the experiments presented in this thesis, the coupling laser is stabilized to the two-photon resonance in a thermal cell [18]. The fluctuations of the two lasers are thus correlated, and consequently the relative linewidth of the two-photon transition can be smaller than the linewidth of the individual lasers.

---

[3] This treatment of laser linewidth is valid providing $\gamma \ll \omega_{\mathrm{p,c}}$ and $\gamma \lesssim \Gamma_e, \Gamma_r$.

The laser-induced dephasing cannot be expressed in the general Lindblad form of Eq. 4.9 as the population terms are unaffected. Instead, a phenomenological operator $\mathcal{L}_d(\sigma)$ is introduced to account for the additional dephasing terms

$$
\mathcal{L}_d(\sigma) = \begin{pmatrix} 0 & -\gamma_p \sigma_{ge} & -\gamma_{rel} \sigma_{gr} \\ -\gamma_p \sigma_{eg} & 0 & -\gamma_c \sigma_{er} \\ -\gamma_{rel} \sigma_{rg} & -\gamma_c \sigma_{re} & 0 \end{pmatrix}.
\tag{4.14}
$$

The OBE equation is then modified as follows

$$
\dot{\sigma} = \frac{i}{\hbar} [\sigma, \mathcal{H}] + \mathcal{L}(\sigma) + \mathcal{L}_d(\sigma).
\tag{4.15}
$$

4.1.2 Steady-State Solution

Probe-Only ($\Omega_c = 0$)

Without the coupling laser, the system reduces to a driven two-level atom. In the absence of spontaneous emission, the population oscillates between states $|g\rangle$ and $|e\rangle$ at a frequency $\Omega = \sqrt{\Omega_p^2 + \Delta_p^2}$, known as Rabi oscillations [3, 19]. The effect of decay from the excited state at rate Γ_e is to damp these Rabi oscillations, causing the system to reach a steady-state on timescales $t \gg \tau_e$.

It is simple to calculate the steady-state of the system by setting the left hand side of Eq. 4.15 to zero and using the normalisation condition $\mathrm{Tr}\{\sigma\} = 1$. This gives the following results for the steady-state populations and coherence terms

$$
\sigma_{ee}^{ss} = (1 - \sigma_{gg}^{ss}) = \frac{1}{2} \frac{\Omega_p^2 \gamma_{eg}}{\gamma_{eg} \Omega_p^2 + \Gamma_e(\gamma_{eg}^2 + \Delta_p^2)},
\tag{4.16a}
$$

$$
\sigma_{eg}^{ss} = (\sigma_{ge}^{ss})^* = \frac{\Omega_p}{2} \frac{\Delta_p - i\gamma_{eg}}{\Omega_p^2/2 + \gamma_{eg}^2 + \Delta_p^2},
\tag{4.16b}
$$

where $\gamma_{eg} = \Gamma_e/2 + \gamma_p$.

Weak-Probe ($\Omega_p \ll \Omega_c, \Gamma_e$)

For the full three-level system it is not possible to solve the coupled-equations analytically. Instead, for the case $\Omega_p \ll \Gamma_e, \Omega_c$, the population can be assumed to remain in the ground-state for all times $\sigma_{gg}^{ss} = 1$. Using this assumption, the steady-state coherence for the probe transition is

$$\sigma_{eg}^{\text{ss}} = -\frac{\mathrm{i}\,\Omega_{\text{p}}/2}{\gamma_{ge} - \mathrm{i}\,\Delta_{\text{p}} + \dfrac{\Omega_{\text{c}}^2/4}{\gamma_{gr} - \mathrm{i}(\Delta_{\text{p}} + \Delta_{\text{c}})}}, \tag{4.17}$$

where $\gamma_{gr} = \Gamma_r/2 + \gamma_{\text{rel}}$.

4.1.3 Complex Susceptibility

The susceptibility at the probe laser frequency ω_{p} for a uniform atomic density of ρ atoms per unit volume is related to the density matrix by [17]

$$\chi(\omega_{\text{p}}) = -\frac{2\rho d_{eg}^2}{\varepsilon_0 \hbar\, \Omega_{\text{p}}}\,\text{Tr}\{\sigma\hat{\pi}^-\} \tag{4.18}$$

$$= -\frac{2\rho d_{eg}^2}{\varepsilon_0 \hbar\, \Omega_{\text{p}}}\,\sigma_{eg}.$$

χ is typically a complex parameter, and can be resolved into the real and imaginary components, $\chi = \chi_{\text{R}} + \mathrm{i}\chi_{\text{I}}$. These components are related by the Kramers-Kronig relations [9]

$$\chi_{\text{R}} = \frac{1}{\pi}\mathcal{P}\int_{-\infty}^{\infty} \frac{\chi_{\text{I}}(\omega')\mathrm{d}\omega'}{\omega' - \omega}, \tag{4.19a}$$

$$\chi_{\text{I}} = -\frac{1}{\pi}\mathcal{P}\int_{-\infty}^{\infty} \frac{\chi_{\text{R}}(\omega')\mathrm{d}\omega'}{\omega' - \omega}, \tag{4.19b}$$

where $\mathcal{P}$ denotes the principle value of the integral. These relations mean that the real part of the susceptibility can be calculated using measurements of the imaginary susceptibility, providing the frequency dependence is known; and vice-versa.

From the steady-state solution of Eq. 4.17, the susceptibility of the three-level system in the weak-probe limit is

$$\chi(\Delta_{\text{p}}) = \frac{\mathrm{i}\rho d_{eg}^2/\varepsilon_0\hbar}{\gamma_{ge} - \mathrm{i}\,\Delta_{\text{p}} + \dfrac{\Omega_{\text{c}}^2/4}{\gamma_{gr} - \mathrm{i}(\Delta_{\text{p}} + \Delta_{\text{c}})}}. \tag{4.20}$$

4.1.4 Optical Response

In an experiment it is not the complex susceptibility that is measured, but the back-action on the probe field propagating through the medium. The optical properties are related to the susceptibility through the refractive index n by

$$n = \sqrt{1 + \chi} \simeq 1 + \frac{\chi_R + i\chi_I}{2}, \tag{4.21}$$

where the approximation is valid providing $|\chi| \ll 1$, valid for the experiments presented in Part II for which $|\chi| \lesssim 10^{-4}$.

For a probe field propagating a distance ℓ through the medium, the output electric field is

$$E = E_0 e^{i(kn\ell - \omega t)} = E_0 e^{-k\chi_I \ell/2} e^{i(k\chi_R \ell/2 - \omega t)}, \tag{4.22}$$

where $k = 2\pi/\lambda$ is the wavevector. The medium can therefore attenuate the field proportional to the imaginary part of the susceptibility, and change the relative phase proportional to the real part of the susceptibility. The resulting phase shift and intensity transmission are given by

$$T = \frac{I}{I_0} = \exp(-k\chi_I \ell), \tag{4.23a}$$

$$\Delta\phi = k\chi_R \ell/2. \tag{4.23b}$$

Thus from measurements of transmission or phase along a known path length it is possible to infer the value of the susceptibility.

4.2 Electromagnetically Induced Transparency

To understand how transparency can arise in the three-level system, it is instructive to diagonalise the Hamiltonian of Eq. 4.6 to obtain the eigenstates on the two-photon resonance ($\Delta = \Delta_p + \Delta_c = 0$), given by [11]

$$|+\rangle = \sin\theta \sin\phi |g\rangle + \cos\phi |e\rangle + \cos\theta \sin\phi |r\rangle, \tag{4.24a}$$

$$|D\rangle = \cos\theta |g\rangle - \sin\theta |r\rangle, \tag{4.24b}$$

$$|-\rangle = \sin\theta \cos\phi |g\rangle - \sin\phi |e\rangle + \cos\theta \cos\phi |r\rangle, \tag{4.24c}$$

where θ and ϕ are the Stückelberg mixing angles defined as

$$\tan\theta = \frac{\Omega_p}{\Omega_c}, \quad \tan 2\phi = \frac{\sqrt{\Omega_p^2 + \Omega_c^2}}{\Delta_p}. \tag{4.25}$$

In the weak probe limit ($\Omega_p \ll \Omega_c, \Gamma_e$), the mixing angle $\theta \to 0$ to give $|\pm\rangle = (|r\rangle \pm |e\rangle)/\sqrt{2}$ and $|D\rangle = |g\rangle$ on resonance ($\Delta_p = 0$). The probe laser only couples to the $|e\rangle$ component of the states $|\pm\rangle$, which have equal magnitude but opposite

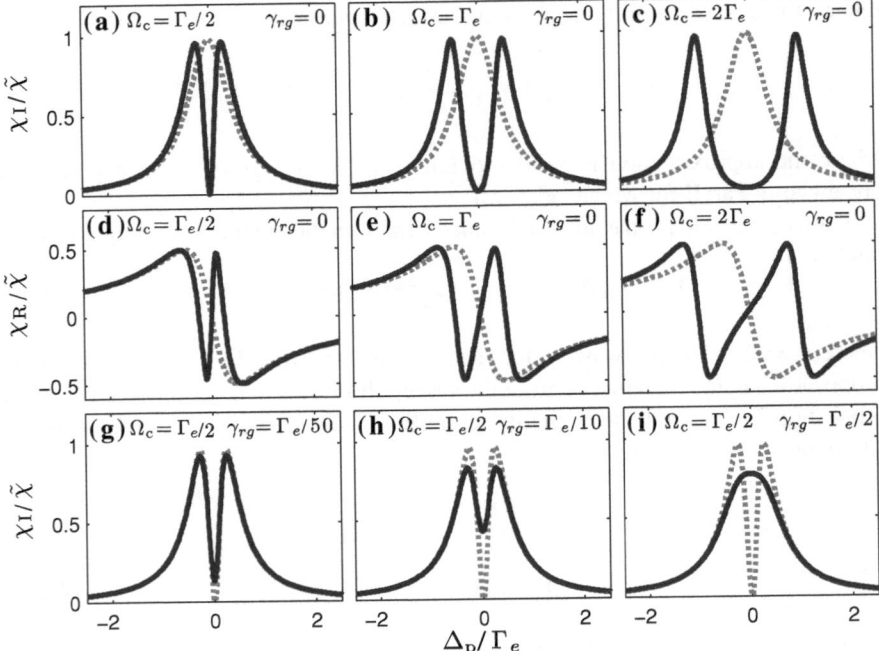

Fig. 4.2 Three-level atom susceptibility. **a–f** show comparisons to the EIT condition (*solid line*) to a two-level atom (*dashed line*), revealing the narrow transparency window on resonance and the associated dispersion feature in χ_R. **g–h** illustrate the effect of a finite laser-linewidth (*solid line*) compared to $\gamma_{rg} = 0$ (*dashed line*), which limits the visibility of the transparency. All curves are calculated for $\Omega_p = \Gamma_e/10$ and scaled relative to the probe-only resonant susceptibilty $\tilde{\chi} = 2\rho d_{eg}^2/\epsilon_0 \hbar \Gamma_e$

signs. The result is a destructive interference of the excitation pathways, so the probe laser is no longer absorbed. The state $|D\rangle$ is therefore known as a *dark state* as it is not coupled to the light field, having a zero-energy eigenvalue. Since states $|\pm\rangle$ include the radiative state $|e\rangle$, they decay to populate $|D\rangle$ on timescales of order τ_e. This coherent phenomenon is known as electromagnetically induced transparency (EIT) [10, 11], as the strong coupling laser changes the optical properties of the medium from resonant absorption of the probe laser to perfect transmission. EIT was first observed experimentally by Boller et al. [20] using a Λ-configuration, where the $|r\rangle$ state is replaced by a second ground-state transition, enabling very narrow resonances.

Figure 4.2 shows the susceptibility for a range of parameters to illustrate the effect of EIT. In (a), the coupling laser can be seen to switch the imaginary susceptibility on resonance (and hence absorption) from a maximum to zero, giving complete transparency assuming $\Gamma_r \to 0$, which gives a resonantly enhanced $\chi^{(3)}$ in the medium [10]. As Ω_c is increased, the EIT resonance splits (known as Autler-Townes splitting [21]), increasing the bandwidth of the transparency. The Kramers-Kronig relations show it is not possible to have a change in χ_I without a concommitant change in χ_R. This can be seen in (d) with the appearance of a steep dispersive feature. The group velocity v_g of light as it passes through the atomic medium is [11]

$$v_g = \frac{c}{n(\omega_p) + \omega_p \dfrac{dn}{d\omega_p}}, \qquad (4.26)$$

which gives a drastically reduced group velocity on resonance due to the gradient of χ_R, leading to light being slowed. Hau et al. used EIT in a BEC to reduce the speed of light to $17\,\mathrm{ms}^{-1}$ [22], corresponding to $\chi^{(3)} = 4.8 \times 10^{-8}\,\mathrm{m}^2\,\mathrm{V}^{-2}$, the largest recorded optical non-linearity in a cold atom system. As well as slowing light, pulses can be stored in the medium for a duration of $1\,\mathrm{ms}$ [23]. This can be used as an optical memory, and single photon storage has been demonstrated between two spatially separate locations [24, 25].

EIT is very sensitive to dephasing, which destroys the coherence of the dark state. In Fig. 4.2g–i the effect of the relative linewidth of the two-photon resonance γ_{rg} is shown. The laser induced dephasing mixes the eigenstates, causing the dark state to gain a contribution from $|e\rangle$ and hence suppression of the transmission on resonance. It is therefore necessary for $\gamma_{rg} \ll \Omega_c, \Gamma_e$ to observe EIT. As well as dephasing, the Doppler effect is important in thermal samples as the velocity averaging can wash-out the transmission on the two-photon resonance [17]. For the ladder system, this can be minimised using counter-propagating probe and coupling lasers, however EIT can only be observed if $k_p < k_c$ [26] unless cold atoms are used.

4.2.1 Related Phenomena

If the probe Rabi frequency is increased beyond the weak-probe limit, the state $|D\rangle$ is given by Eq. 4.24b, forming a super-position of states $|g\rangle$ and $|r\rangle$. For $\Gamma_r \to 0$, this remains a dark state and population is transferred into $|r\rangle$ without population of the radiative $|e\rangle$ state. This is known as coherent population trapping (CPT) [27–29], illustrated in Fig. 4.3 which shows the evolution of population with the ratio of Ω_p to Ω_c (and hence θ). An important distinction between EIT and CPT is that EIT only occurs in an optically thick medium, where the atomic coherences induced by the lasers cause a back-action on the probe laser.

In CPT the system is prepared in the dark state by decay from $|\pm\rangle$, limiting the fidelity of the state preparation for $\theta > 0$ [30]. Alternatively, an adiabatic evolution of the field using a counter-intuitive pulse sequence allows smooth evolution from $\theta = 0$ with all atoms in $|g\rangle$ to $\theta = \pi/2$ with all atoms in $|r\rangle$. This is known as stimulated Raman adiabatic passage (STIRAP) [31, 32] and can be used to transfer population via the dark state with almost 100 % efficiency.

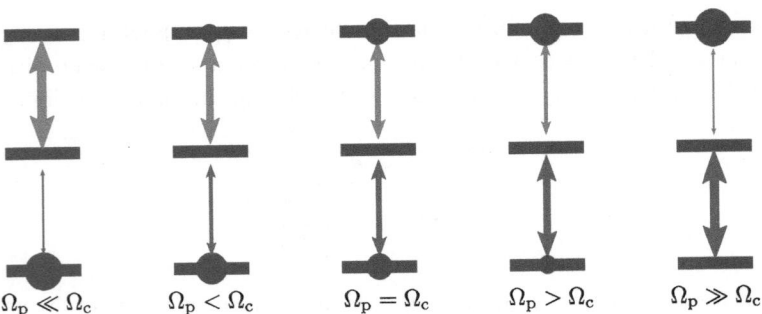

$$\Omega_p \ll \Omega_c \qquad \Omega_p < \Omega_c \qquad \Omega_p = \Omega_c \qquad \Omega_p > \Omega_c \qquad \Omega_p \gg \Omega_c$$

Fig. 4.3 Dark-state populations as a function of Ω_p, Ω_c. On the *left-hand* side $\Omega_p \ll \Omega_c$, resulting in electromagnetically induced transparency (EIT). On the *right-hand* side $\Omega_p \gg \Omega_p$, leading to coherent population transfer (CPT) into $|r\rangle$. If the laser intensities are changed in time from *left* to *right*, this is equivalent to STIRAP

4.3 Summary

The evolution of the three-level system can be calculated using the optical Bloch equations to model the effects of spontaneous emission and laser linewidth. This enables the density matrix to be known for any time t, and hence the optical properties of the medium from calculation of the complex susceptibility.

On the two-photon resonance the lasers drive the system into a coherent dark state $|D\rangle$, which is not coupled to the probe field. For $\Omega_p \ll \Omega_c$, this dark state corresponds to a narrow transmission window in the absorption feature of $|e\rangle$, leading to the phenomena of EIT. The associated change in refractive index creates a steep dispersive feature which can be used to slow and store light in the medium. Thus EIT provides a means to create large optical non-linearities without a significant absorption on resonance, as is the case for a two-level atom. Another important feature of EIT is that in this weak probe regime $|D\rangle \equiv |g\rangle$, allowing the Rydberg state to be probed without transferring population into the state. However, as the probe power is increased atoms are excited to the Rydberg states and it is necessary to consider the effects of the strong dipole-dipole interactions discussed in the previous chapter.

References

1. L. Allen, J.H. Eberly, *Optical resonance and two-level atoms* (Dover Publications, New York, 1987)
2. C. Cohen-Tannoudji, J. Dupont-Roc, G. Grynberg, *Atom-Photon Interactions* (Wiley-Interscience, New York, 1998)
3. I.I. Rabi, Space quantization in a gyrating magnetic field. Phys. Rev. **51**(8), 652 (1937)
4. A. Ashkin, Trapping of atoms by resonance radiation pressure. Phys. Rev. Lett. **40**(12), 729 (1978)

5. R. Grimm, M. Weidemuller, Y.B. Ovchinnikov, Optical dipole trap for neutral atoms. Adv. At. Mol. Opt. Phys. **42**(95), 170 (2000)
6. T.W. Hänsch, A.L. Schawlow, Cooling of gases by laser radiation. Opt. Commun. **13**(1), 68 (1975)
7. D. Wineland, H. Dehmelt, Proposed $10^{14} \Delta\nu < \nu$ laser fluoresence spectroscopy on Ti$^+$ mono-Ion oscillator III. Bull. Am. Phys. Soc. **20**, 637 (1975)
8. C.S. Adams, E. Riis, Laser Cooling and trapping of neutral atoms. Prog. Quantum Electron. **21**(1), 1 (1997)
9. R.W. Boyd, *Nonlinear Optics*, 3rd edn. (Academic Press, USA, 2008)
10. S.E. Harris, J.E. Field, A. Imamoğlu, Nonlinear optical processes using electromagnetically induced transparency. Phys. Rev. Lett. **64**(10), 1107 (1990)
11. M. Fleischhauer, A. Imamoglu, J. Marangos, Electromagnetically induced transparency: Optics in coherent media. Rev. Mod. Phys. **77**, 633 (2005)
12. J. Dalibard, Y. Castin, K. Mølmer, Wave-function approach to dissipative processes in quantum optics. Phys. Rev. Lett. **68**(5), 580 (1992)
13. G. Lindblad, On the generators of quantum dynamical semigroups. Commun. Math. Phys. **48**(2), 119 (1976)
14. K. Mølmer, Y. Castin, Monte Carlo wavefunctions in quantum optics. Quantum Semiclass. Opt. **8**(1), 49 (1996)
15. W.T. Silfvast, *Laser Fundamentals* (CUP, New York, 1996)
16. S. Sultana, M. Suhail Zubairy, Effect of finite bandwidth on refractive-index enhancement and lasing without inversion. Phys. Rev. A **49**(1), 438 (1994)
17. J. Gea-Banacloche, Y. Li, S. Jin, M. Xiao, Electromagnetically induced transparancy in ladder-type inhomogeneously broadened media: Theory and experiments. Phys. Rev. A **51**(1), 576 (1995)
18. R.P. Abel, A.K. Mohapatra, M.G. Bason, J.D. Pritchard, K.J. Weatherill, U. Raitzsch, C.S. Adams, Laser frequency stabilization to excited state transitions using electromagnetically induced transparency in a cascade system. Appl. Phys. Lett. **94**(7), 071107 (2009)
19. C.J. Foot, *Atomic Physics* (OUP, Oxford, 2005)
20. K.-J. Boller, A. Imamoğlu, S.E. Harris, Observation of electromagnetically induced transparency. Phys. Rev. Lett. **66**(20), 2593 (1991)
21. S.H. Autler, C.H. Townes, Stark effect in rapidly varying fields. Phys. Rev. **100**(2), 703 (1955)
22. L.V. Hau, S.E. Harris, Z. Dutton, C.H. Behroozi, Light Speed reduction to 17 metres per second in an ultracold atomic gas. Nature **397**, 594 (1999)
23. C. Liu, Z. Dutton, C.H. Behroozi, L.V. Hau, Observation of coherent optical information storage in an atomic medium using halted light pulses. Nature **409**, 490 (2001)
24. T. Chaneliàre, D.N. Matsukevich, S.D. Jenkins, S.-Y. Lan, T.A.B. Kennedy, A. Kuzmich, Storage and retrieval of single photons transmitted between remote quantum memories. Nature **438**, 833 (2005)
25. M.D. Eisaman, A. André, F. Massou, M. Fleischhauer, A.S. Zibrov, M.D. Lukin, Electromagnetically induced transparency with tunable single-photon pulses. Nature **438**, 837 (2005)
26. J.R. Boon, E. Zekou, D. McGloin, M.H. Dunn, Comparison of wavelength dependence in cascade-, Λ-, and Vee-type schemes for electromagnetically induced transparency. Phys. Rev. A **59**(6), 4675 (1999)
27. G. Alzetta, A. Gozzini, L. Moi, G. Orriols. An experimental method for the observation of R.F. transitions of laser beat resonances in oriented Na vapour. Il Nuovo Cimento B. **36**(1) 5, (1976)
28. E. Arimondo, G. Orriols, Nonabsorbing atomic coherences by coherent two-Photon transitions in a three-level optical pumping. Lettere Al Nuovo Cimento **17**(10), 333 (1976)
29. E. Arimondo, Coherent population trapping in laser spectroscopy. Prog. Opt. **35**, 257 (1996)
30. E. Arimondo, Relaxation processes in coherent-population trapping. Phys. Rev. A **54**(3), 2216 (1996)
31. J. Oreg, F.T. Hioe, J.H. Eberly, Adiabatic following in multilevel systems. Phys. Rev. A **29**(2), 690 (1984)
32. K. Bergmann, H. Theuer, B.W. Shore, Coherent population transfer among quantum states of atoms and molecules. Rev. Mod. Phys. **70**(3), 1003 (1998)

Chapter 5
Cooperative Phenomena

In the previous chapter the atom-light interaction for a single atom was assumed to describe the behaviour of a macroscopic sample, calculating the susceptibility of a uniform gas with density ρ using the single atom density matrix. This description is valid providing the atoms are both independent and identical. In some circumstances the atoms behave independently but the overall response of the system depends on the sum over all atoms. This is known as collective behaviour. An example is spin-echo, where each atom dephases at a different rate but reversing the phase leads to a restoration of the initial state, resulting in a collective emission.

If the atoms are no longer independent but instead correlated, the ensemble properties become fundamentally different to those of an isolated atom, and cannot be determined by summing over the individual responses. This situation is described by Mandel and Wolf in Sect. 16 of Ref. [1],

> In other cases it is essential to include the effect of each atom on all the other atoms, because this modifies the behaviour of each in a significant way. These phenomena, such as self induced transparency and superradiance are collective effects in a deeper sense. They are sometimes called cooperative effects.

Cooperative phenomena therefore occur in systems when the atom-atom interactions cannot be treated simply as a perturbation, but instead dominate the evolution of the ensemble. The main mechanism for such effects is the interaction of each atom with the dipole-radiation field of the surrounding atoms, resulting in a well defined phase between the dipoles that modifies the optical properties.

5.1 Cooperative Behaviour for Two Atoms

To see how cooperative behaviour can arise, it is instructive to first study the case of a pair of atoms. Consider an atom located at the origin which has dipole moment d oscillating at frequency ω. The electric field at position R due to this dipole is given by [2]

J. D. Pritchard, *Cooperative Optical Non-Linearity in a Blockaded Rydberg Ensemble*, 49
Springer Theses, DOI: 10.1007/978-3-642-29712-0_5,
© Springer-Verlag Berlin Heidelberg 2012

$$E(R) = \frac{1}{4\pi\epsilon_0} \left\{ \frac{k^2}{R} (\hat{R} \times d) \times \hat{R} + \left(\frac{1}{R^3} - \frac{ik}{R^2} \right) [3\hat{R}(\hat{R} \cdot d) - d] \right\} e^{-ikR}, \quad (5.1)$$

where $R = |R|$, $\hat{R} = R/R$ and $k = \omega/c$ is the wavevector. The e^{-ikR} phase term arises due to the finite speed of light c, which creates a retarded potential that lags behind in time by a factor $t - R/c$.

The electric field can be expressed in two distinct limits:

Far-field $(kR \gg 1)$: At long range the $1/R$ term dominates,

$$E(R) = \frac{k^2 (\hat{R} \times d) \times \hat{R}}{4\pi\epsilon_0 R} e^{-ikR}, \quad (5.2)$$

which describes the propagation of a transverse spherical wave; this limit is also known as the radiation zone.

Near-field $(kR \lesssim 1)$: In the near-field the higher-order terms dominate, equivalent to the field of a static dipole [3].

If we now place a second atom at position R which is at an angle θ to the z-axis (Fig. 3.1a), it will experience an interaction with the dipole electric field of the first atom that is described by the Hamiltonian $\mathcal{H}_{dd} = -d \cdot E(R)$. Using the definition of the spontaneous decay rate Γ from the excited state in terms of the dipole moment [4]

$$\Gamma = \frac{k^3 d^2}{3\hbar\pi\epsilon_0}, \quad (5.3)$$

this interaction Hamiltonian can be written in terms of Γ as

$$\mathcal{H}_{dd} = -\frac{3\hbar\Gamma}{4} \left[\frac{1}{kR} \sin^2\theta + \left(\frac{1}{(kR)^3} - \frac{i}{(kR)^2} \right) (3\cos^2\theta - 1) \right] e^{-ikR}. \quad (5.4)$$

If the atoms are initially excited to state $|ee\rangle$, they will decay into a Dicke state $|\pm\rangle = (1/\sqrt{2})(|eg\rangle \pm |ge\rangle)$ [5] by the spontaneous emission of a single photon. This first emission creates a phase between the dipoles as either of the atoms could have decayed. It can be shown that the effect of the coupling between the two atoms due to $\mathcal{H}_{dd}$ is to modify both the decay rate and the energy of the states $|\pm\rangle$ to give [6]

$$\Gamma_\pm = \Gamma \pm \Gamma_{12}, \quad (5.5a)$$
$$W_\pm = \hbar\omega_0 \pm \hbar\Delta_{12}, \quad (5.5b)$$

where Γ_{12} is the enhanced broadening term and Δ_{12} is the shift in the energy of the $|\pm\rangle$ states. These are related to the real and imaginary parts of $\mathcal{H}_{dd}$ as follows[1] [6] (C.S. Adams, Private Communication, 2010)

---

[1] Comparison of $\mathcal{H}_{dd}$ to D_{12} in Eq. (9) of [6] yields $D_{12} \equiv 2i\mathcal{H}_{dd}/\hbar\Gamma$.

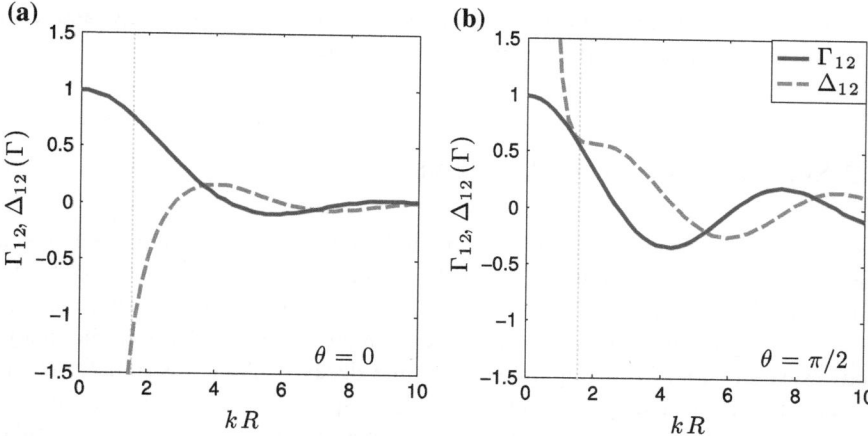

Fig. 5.1 Dipole–dipole induced broadening Γ_{12} and level-shift Δ_{12} for a pair of atoms. The level shift dominates over dephasing for $kR < \pi/2$, indicated by the *dashed line*

$$\Gamma_{12} = -2\text{Im}\{\mathscr{H}_{\text{dd}}\}/\hbar, \tag{5.6a}$$

$$\Delta_{12} = 2\text{Re}\{\mathscr{H}_{\text{dd}}\}/\hbar. \tag{5.6b}$$

Figure 5.1 plots these quantities as a function of kR for two different alignment configurations, showing that it is possible to observe decay of the single excitation from $|+\rangle$ at a maximum rate of 2Γ. This is known as a *superradiant* state, as the spontaneous emission occurs faster than Γ. Conversely, the $|-\rangle$ state is a *sub-radiant* state, as the decay rate at close separations is less than Γ, corresponding to an enhanced lifetime.

An important feature of Fig. 5.1 is the relative magnitude of the energy shift (Δ_{12}) compared to the broadening rate (Γ_{12}). In the region $\pi/2 \lesssim kR < 10$, the dominant effect is the modification of the decay rate to give sub- or superradiant decay. This effect has been demonstrated experimentally for a pair of trapped ions at variable separations [7]. However, at shorter range ($kR < \pi/2$) the energy-shift diverges. Using Eqs. 5.1 and 5.6b, the energy shift is given by

$$\Delta_{12} = \frac{\boldsymbol{d} \cdot \boldsymbol{d} - 3(\hat{\boldsymbol{R}} \cdot \boldsymbol{d})(\hat{\boldsymbol{R}} \cdot \boldsymbol{d})}{4\pi\epsilon_0 R^3}. \tag{5.7}$$

This is equivalent to the dipole–dipole interaction between two static dipoles of Eq. 3.1 that was used to calculate the interaction strength for a pair of Rydberg atoms in Chap. 3.

In summary, the dipole–dipole coupling between two atoms significantly modifies the effective energy and lifetime of the pair states. The effect cannot be reproduced by considering each atom as a single independent emitter. Only by including the correlations between the dipoles of each atom can the superradiant decay and level shift can be explained, making the process cooperative.

5.2 Superradiance from $\mathcal{N}$-Atoms

Extension from a pair of atoms to a system of $\mathcal{N}$-atoms is a complex problem which has received much attention [5, 8–10]. In the original paper on the topic, Dicke [5] assumed an ensemble of $\mathcal{N}$-atoms were localised to a region small compared to λ, allowing the e^{ikR_i} phase-factors of each atom to tend to unity in the many-body wavefunction. In this approximation, the decay of $\mathcal{N}$-atoms initially in the excited state $|e\rangle$ is analogous to an ensemble of spin-1/2 particles precessing in a magnetic field. The system starts in a fully symmetric state $|J, M\rangle$ with total angular momentum $J = \mathcal{N}/2$, which has a projection along z of $M = \mathcal{N}/2$. As each photon is emitted, the system decays from $M \to M - 1$ at a rate given by [10]

$$\Gamma_{M \to M-1} = (J + M)(J - M + 1)\Gamma. \tag{5.8}$$

The initial decay from $|J, J\rangle$ to $|J, J - 1\rangle$ occurs at rate $\mathcal{N}\Gamma$, the expected decay rate for $\mathcal{N}$-atoms. This projects the system into a symmetric superposition state as any one of the $\mathcal{N}$ atoms could have decayed, introducing correlations between the dipoles in the system. As subsequent photons are emitted, these correlations cause the rate of spontaneous emission to increase, reaching a maximum value of approximately $\mathcal{N}^2/4\Gamma$ for the decay of $|J, 0\rangle$ when exactly half the atoms have decayed. This is shown schematically in Fig. 5.2a.

To illustrate the dynamic effects of this enhanced dephasing, Fig. 5.2b shows the decay of the average magnetisation $\langle M(t)\rangle$ for $\mathcal{N}=10$, compared to the decay of $\mathcal{N}$ independent atoms. This reveals an initial delay followed by rapid decay at a rate much faster than Γ. The intensity of the emission is related to $I \propto -d\langle M(t)\rangle/dt$ [10], which is plotted in Fig. 5.2c. The superradiant decay is observed as a pulsed emission, with a peak intensity proportional to $\mathcal{N}^2$ as opposed to an exponentially decaying intensity expected for independent atoms. The characteristic delay time for the emission is $\tau_d \sim \ln(\mathcal{N})/\mathcal{N}\Gamma \ll 1/\Gamma$ [9], getting increasingly shorter as more atoms are included.

For a sample of finite size, it is necessary to include the e^{ikR_i} phase factors. This creates a phase-matching condition in the sample, resulting in an angular emission pattern that is strongly dependent upon the sample geometry [9]. These phases can also break the symmetry assumed in the simple spin-model above, projecting the system into an effective sub-space with $J < \mathcal{N}/2$, suppressing the superradiant emission. This finite size effect can be accounted for by replacing Γ with $\mathcal{C}\Gamma$ where $\mathcal{C}$ is the *cooperativity parameter*. For a spherical volume with radius R_0, this is given by [11]

$$\mathcal{C} = \frac{9 \left[\sin(kR_0) - kR_0 \cos(kR_0)\right]^2}{(kR_0)^6}. \tag{5.9}$$

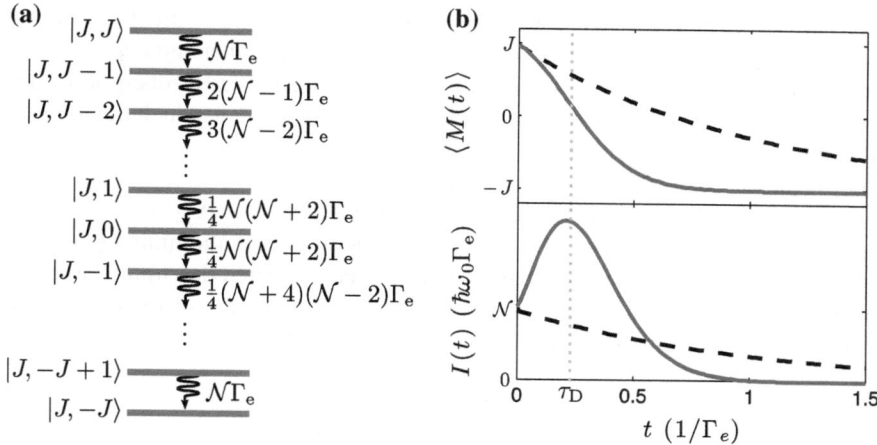

Fig. 5.2 Dicke model for superradiant emission. **a** Decay rate between symmetric states $|J, M\rangle$ occurs at rate $(J+M)(J-M-1)\Gamma$, reaching a peak of $\sim\!\mathcal{N}^2/4$ at $M = 0$; **b** average magnetisation $\langle M(t)\rangle$ and **c** radiated intensity $I(t)$ calculated for $\mathcal{N} = 10$ atoms (*solid*) compared to emission from independent atoms (*dashed*), revealing a pulsed emission with a characteristic delay τ_D marked as a *vertical line*

For the experiments performed on laser cooled atoms in a magneto-optical trap (MOT) the sample size is around 1 mm, for which $\mathcal{C}$ becomes negligible for $n < 20$.

Another important effect in finite size samples is the distribution of level shifts in the system due to variation in the separations R_{ij} between atoms. This causes the relative phases between the dipoles to evolve at different rates, which destroys the coherence built up during the collective emission, turning off the superradiance. This effect is known as *van der Waals dephasing* which is analogous to an inhomogeneous dephasing of the dipoles [10].

To observe these cooperative effects it is necessary to localise atoms to dimensions of λ. For optical transitions, this condition is very challenging, as at these short ranges alternative dephasing processes such as collisional broadening destroy the coherences between neighbouring dipoles, suppressing the superradiant emission. Early observations of superradiance were therefore for transitions in the infrared with $\lambda \sim 100\,\mu$m in HF molecules [12] and 2–9 μm in Na atoms [13].

Rydberg atoms, however, have transitions to close-lying n states that are in the millimeter or microwave region, so even at modest densities it is possible to observe superradiance, first demonstrated by Gounand et al. [14] for the $12S_{1/2}$ state in Rb. Superradiance from the original Rydberg state leads to significant population transfer into close-lying states, which may then also undergo superradiant decay. This is known as a superradiant cascade, which is typically detected indirectly through the distribution of population over a range of $n\ell$ states rather than detection of the emitted field. Recent experiments exploring superradiant cascade from ultracold gases have demonstrated good agreement between calculated and observed population dynamics [11, 15].

For the low-ℓ Rydberg states, the relative strengths of the broadening Γ_{12} and shift Δ_{12} from Eq. 5.6 can be considered in terms of the principal quantum number using the fact that the spontaneous decay rate, $\Gamma \propto \omega^3 d^2$. For an isolated Rydberg atom, this is dominated by the high-frequency coupling to the ground-state which has a dipole moment $d \propto n^{*-3/2}$, leading to $\Gamma \propto n^{*-3}$. For superradiance however, the relevant decay channel is that of the close-lying states with $\omega \propto n^{*-3}$ and $d \propto n^{*2}$, and hence $\Gamma \propto n^{*-5}$. Conversely, the dipole–dipole energy shift scales as $\Delta_{12} \propto d^2 \propto n^{*4}$, so combining these scalings with the fact that Δ_{12} dominates in the limit $kR < \pi/2$, for high n states the superradiant broadening can be neglected due to the large van der Waals dephasing. Therefore, for large n^*, the dipole–dipole interactions can be treated purely as an energy shift of the multiply excited Rydberg states.

5.3 Dipole Blockade

In this high n^* limit, where the dipole–dipole interactions can be treated as an energy shift, this leads to a pair of atoms excited to the Rydberg state $|rr\rangle$ experiencing an interaction energy $V(R)$, as discussed in Chap. 3. At large separations the interaction can be neglected, and the atoms behave independently. For the case of a pair of atoms resonantly excited from $|g\rangle$ to $|r\rangle$ at Rabi frequency Ω, shown schematically in Fig. 5.3a, the atoms will populate $|rr\rangle$ at a rate Ω. If the atoms are now moved closer together, the interaction causes the $|rr\rangle$ state to be detuned from resonance with the laser, preventing excitation of the $|rr\rangle$ state. This process is known as *dipole blockade* [16], and occurs when the interaction shift is larger than the linewidth of the $|rr\rangle$ state

$$V(R) > \hbar \times \max(\Omega, \Gamma_r), \tag{5.10}$$

where the linewidth of the $|rr\rangle$ state is determined by the larger of the natural linewidth Γ_r or the power-broadened width Ω. As Rydberg states are relatively long-lived, typically $\Omega \gg \Gamma_r$ for experimental parameters.

The condition $V(R_b) = \hbar\Omega$ defines the blockade radius R_b, which for van der Waals interactions $V(R) = C_6/R^6$ is given by

$$R_b = \sqrt[6]{\frac{C_6}{\Omega}}, \tag{5.11}$$

forming a sphere around the Rydberg atom in which only a single Rydberg excitation is allowed. The blockade mechanism is important as it enables deterministic creation of singly-excited entangled states, which can be used for implementing quantum gates [16–19].

For an ensemble of $\mathcal{N}$-atoms localised within a radius $R < R_b$, the blockade mechanism projects the system into the symmetric entangled state

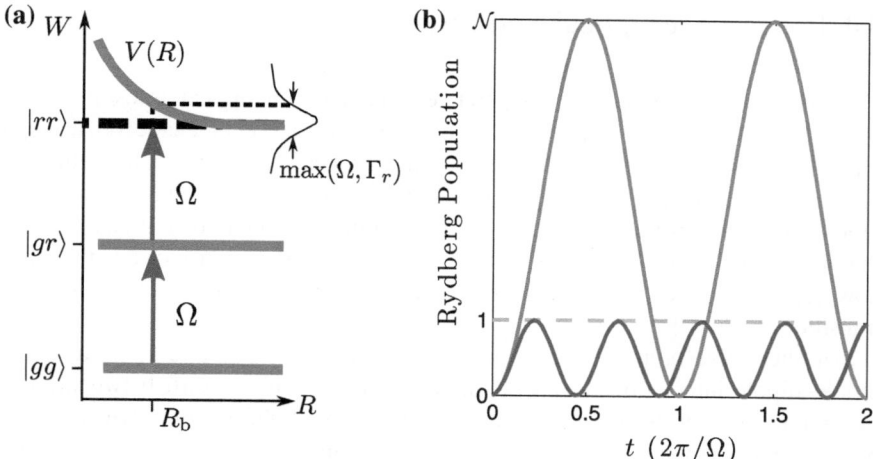

Fig. 5.3 Rydberg dipole blockade. **a** Dipole–dipole interactions shift the energy of state $|rr\rangle$ by $V(R)$, detuning the pair-state from resonance with the excitation laser. If $V(R) > \hbar \times \max(\Omega, \Gamma_r)$ then $|rr\rangle$ cannot be populated, known as dipole-blockade; **b** for $\mathcal{N}$ independent atoms the Rydberg population Rabi flops between 0 and $\mathcal{N}$ with frequency Ω. Dipole blockade causes oscillations to a collective state with a single excitation with an enhanced frequency $\sqrt{\mathcal{N}}\Omega$

$$|g^{\mathcal{N}-1}r\rangle = \frac{1}{\sqrt{\mathcal{N}}} \sum_{i=1}^{\mathcal{N}} e^{-i\mathbf{k}\cdot\mathbf{R}_i}|g_1, g_2, \ldots, r_i, \ldots, g_{\mathcal{N}}\rangle, \qquad (5.12)$$

as each of the atoms are equally likely to be excited. The dipole matrix element between $|g^{\mathcal{N}}\rangle$ and $|g^{\mathcal{N}-1}r\rangle$ is now enhanced to give a collective Rabi-frequency $\sqrt{\mathcal{N}}\Omega$, instead of the oscillations between 0 and $\mathcal{N}$ at rate Ω for the non-interacting case. This is illustrated in Fig. 5.3b. If this collective state can be mapped onto an intermediate excited state the result is cooperative emission of a single photon [20], allowing enhanced atom-light coupling for communication of quantum information between atomic ensembles [21].

Early evidence for dipole-blockade in cold atomic samples came indirectly through a saturation in the resonant excitation of high n states [22–29]. The collective scalings have since been seen from measurements on the coherence of a blockaded Rydberg gas [30–32]. More recently, two groups have demonstrated the collective scaling for a pair of atoms [33, 34], as well as realising entanglement [35] and performing a C-NOT gate [36] using two single atoms loaded in microscopic dipole traps with an atomic separation of $R \sim 3\,\mu$m.

5.4 Cooperative Optical Non-Linearity

The challenges of observing cooperative behaviour on optical transitions with $k = 2\pi/\lambda_{opt}$ make it impossible to directly use dipole blockade on a single photon transition to realise non-linear photonic devices. However, this limitation can be overcome by mapping the large dipole–dipole interactions from the microwave dipoles of the Rydberg states with $k' = 2\pi/\lambda_\mu$ onto a strong optical transition from the ground state using EIT. The result is that the optical response of a single atom now depends on the surrounding atoms even though $kR \gg 1$, as for the Rydberg transitions $k'R \ll 1$.

An alternative proposal has been put forward by Friedler et al. [37] to utilise dipole–dipole interactions to create an accumulated π phase-shift between a pair of photons counter-propagating through an EIT slow-light medium, which can be realised using atoms loaded into a hollow-core fibre [38]. In such a system the interactions can be treated as a perturbation on the propagation of the photons through the medium. However, in the following it is the cooperative nature of the dipole blockade that gives rise to a non-linear optical response.

5.4.1 N-Atom Model

To explore the effect of dipole blockade on the optical properties of the medium, it is necessary to develop a many-atom model of the three-level EIT system introduced in Sect. 4.1, where the dipole–dipole interactions are included as pair-wise couplings between the Rydberg states of each atom, illustrated in Fig. 5.4. For a system of $\mathcal{N}$-atoms, the wavefunction in the coupled basis is given by

$$|\Psi_\mathcal{N}\rangle = \bigotimes_i^\mathcal{N} |\psi\rangle_i, \tag{5.13}$$

where $|\psi\rangle_i$ is the wavefunction of the ith atom. In this basis, the operator $\hat{O}^{(i)}$ acting on atom i can be expressed in terms of the corresponding operator $\hat{O}$ in the single-atom basis using

$$\hat{O}^{(i)} = I_3^{\otimes(i-1)} \otimes \hat{O} \otimes I_3^{\otimes(\mathcal{N}-1)} \tag{5.14}$$

where I_3 is the rank-3 identity matrix. The Hamiltonian acting on the system is

$$\hat{\mathscr{H}}_\mathcal{N} = \sum_i^\mathcal{N} \hat{\mathscr{H}}^{(i)} + \sum_{i<j}^\mathcal{N} V(R_{ij}) \hat{P}_{rr}^{(i)} \hat{P}_{rr}^{(j)}, \tag{5.15}$$

Fig. 5.4 Schematic of an $\mathcal{N}$-atom system with dipole–dipole coupling between the Rydberg states creating an interaction $V(R_{ij})$ dependent upon the atomic separation

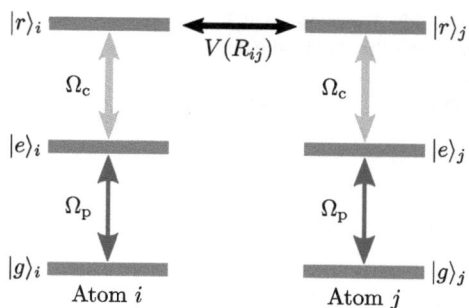

where $\hat{\mathcal{H}}^{(i)}$ is the Hamiltonian for the atom-light coupling of atom i given by Eq. 4.6, $\hat{P}_{rr}^{(i)} = |r\rangle_{ii}\langle r|$ is the projector onto the Rydberg state of atom i and $R_{ij} = |\boldsymbol{R}_i - \boldsymbol{R}_j|$ is the interatomic spacing between atoms i and j.

The time-evolution of the density matrix $\sigma = |\Psi_{\mathcal{N}}\rangle\langle\Psi_{\mathcal{N}}|$ is calculated by solving the optical Bloch equations of Eq. 4.15, however the Lindblad operator now includes a sum over the decay channels m for each atom i,

$$\mathcal{L}(\sigma) = -\frac{1}{2}\sum_{i,m}\left(C_m^{(i)\dagger}C_m^{(i)}\sigma + \sigma C_m^{(i)\dagger}C_m^{(i)}\right) + \sum_{i,m} C_m^{(i)}\sigma C_m^{(i)\dagger}, \tag{5.16}$$

where the operators $C_m^{(i)}$ can be obtained from the operators in Eq. 5.9. Similarly, the operator $\mathcal{L}_d$ for the laser-induced dephasing is modified to include the dephasing of each atom. As this dephasing term is not of the Lindblad form, it can only be generalised with the use of a Hadamard product,[2]

$$\mathcal{L}_d(\sigma) = -\gamma \circ \sigma, \tag{5.17}$$

where the matrix γ contains the contributions from the linewidth of the lasers, defined as

$$\gamma = \sum_i^{\mathcal{N}} J_3^{\otimes(i-1)} \otimes \begin{pmatrix} 0 & \gamma_p & \gamma_{rel} \\ \gamma_p & 0 & \gamma_c \\ \gamma_{rel} & \gamma_c & 0 \end{pmatrix} \otimes J_3^{\otimes(\mathcal{N}-1)}, \tag{5.18}$$

where J_3 is the rank-3 unit matrix (matrix of ones).

Combining these equations together the density matrix can be propagated in time, from which the complex susceptibility at the probe frequency ω_p is obtained by taking the trace over the dipole operators of all the atoms in the system,

---

[2] The Hadamard product defines element-wise multiplication of matrices A and B such that $[A \circ B]_{i,j} = [A]_{i,j} \cdot [B]_{i,j}$ (see e.g.[39], p. 205).

Fig. 5.5 Two-atom model, showing the dipole–dipole interaction $V(R)$ acts as a detuning on the $|rr\rangle$ state

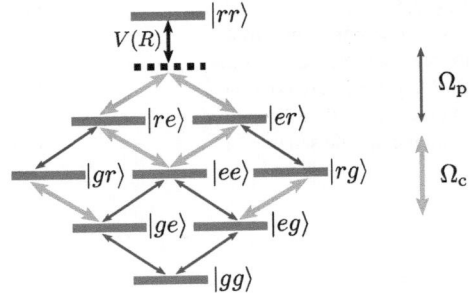

$$\chi(\omega_{\mathrm{p}}) = -\frac{2\rho d_{\mathrm{eg}}^2}{\varepsilon_0 \hbar \, \Omega_{\mathrm{p}}} \mathrm{Tr} \left\{ \sigma \sum_i^{\mathcal{N}} \hat{\pi}^{-(i)} \right\}. \tag{5.19}$$

5.4.2 Two-Atom Model

The simplest system to consider is the case of two-atoms, shown in Fig. 5.5. For large R the interactions can be neglected, and on resonance each of the atoms evolve into the single-atom dark state $|D\rangle$ of Eq. 4.24b, resulting in the product state

$$\begin{aligned} |D^2\rangle = |D\rangle_1 \otimes |D\rangle_2 &= (\cos\theta |g\rangle_1 - \sin\theta |r\rangle_1) \otimes (\cos\theta |g\rangle_2 - \sin\theta |r\rangle_2) \\ &= \cos^2\theta |gg\rangle - \sin\theta\cos\theta(|gr\rangle + |rg\rangle) + \sin^2\theta |rr\rangle, \end{aligned} \tag{5.20}$$

which is independent of the intermediate state and corresponds to perfect transparency on resonance. As the atoms move closer together, the dipole–dipole interactions act to detune the $|rr\rangle$ state by energy $V(R)$, which modifies the dark state. For $V(R) > \gamma_{\mathrm{EIT}}$, where γ_{EIT} is the width of the EIT resonance, the $|rr\rangle$ state is blockaded. Diagonalisation of $\mathcal{H}_{\mathcal{N}}$ gives a new zero-energy eigenstate [18]

$$|\Psi\rangle = \frac{(\cos^2\theta - \sin^2\theta)|gg\rangle - \sin\theta\cos\theta(|gr\rangle + |rg\rangle) + \sin^2\theta |ee\rangle}{\sqrt{\cos^4\theta + 2\sin^4\theta}}. \tag{5.21}$$

This new eigenvector is no longer a simple product state, but instead represents an entangled state where the intermediate pair state $|ee\rangle$ is admixed in place of the Rydberg pair-state. Recalling that $\tan\theta = \Omega_{\mathrm{p}}/\Omega_{\mathrm{c}}$, in the weak probe limit $|\Psi\rangle = |gg\rangle$ which is equivalent to $|D^2\rangle$, resulting in transparency on resonance. As Ω_{p} is increased however, the relative contribution of $|ee\rangle$ increases, which resonantly couples to the probe laser. Thus $|\Psi\rangle$ is no longer a dark state outside of the weak-probe limit.

This eigenstate picture neglects the effect of the radiative decay of state $|e\rangle$, however the true steady-state of the medium can be obtained by solving the optical

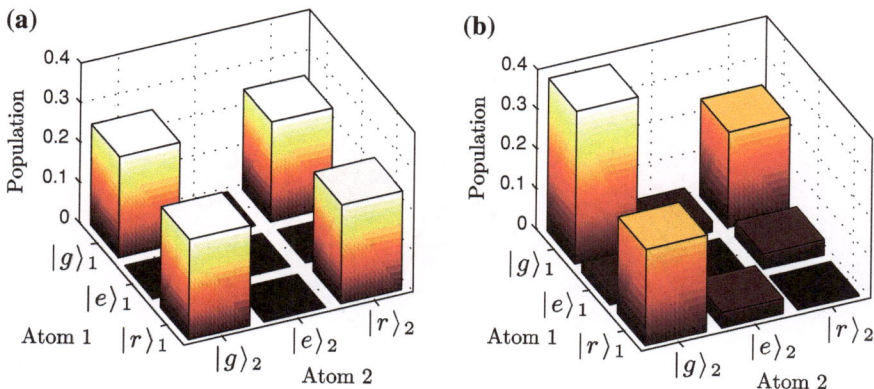

Fig. 5.6 Two-atom dark state populations calculated for $\Omega_p = \Omega_c = \Gamma_e/2$. **a** Non-interacting case ($V = 0$) gives a product state with each atom in the dark state $|D\rangle$; **b** strong interactions ($V = 2\Gamma_e$) blockade population of $|rr\rangle$, causing population of the radiative $|e\rangle$ state. Note unlike the modified eigenstate of 5.21, $|ee\rangle$ is not populated

Bloch equations on the two-photon resonance and extracting the populations from the diagonal elements of the density matrix. Figure 5.6 shows the steady-state solutions calculated using parameters $\Omega_p = \Omega_c = \Gamma_e/2$ ($\theta = \pi/4$) for (a) $V(R) = 0$ and (b) $V(R) = 2\Gamma_e$. In the non-interacting case, the system evolves into the dark state $|D^2\rangle$ with equal population of $|g\rangle$ and $|r\rangle$ due to the choice of mixing angle. In (b) the blockade effect is evident, as there is no population of state $|rr\rangle$. Furthermore, there is also no population in $|ee\rangle$ as predicted by $|\Psi\rangle$. This is because $|ee\rangle$ decays rapidly, leaking population into states $|eg\rangle$, $|ge\rangle$, $|er\rangle$ and $|re\rangle$ which each have approximately 5% population for these parameters. The interpretation of this state is as follows; if one of the atoms is excited into the $|g\rangle - |r\rangle$ dark state, then the other atom has its $|r\rangle$ state detuned by $V(R) > \Omega_c$, meaning it no longer sees the coupling laser. Instead, it now resonantly couples to the probe laser and cycles between states $|g\rangle$ and $|e\rangle$, resulting in population of $|gg\rangle$, $|gr\rangle$, $|ge\rangle$ and $|re\rangle$. As either atom can be excited to the Rydberg state, this leads to an entangled state which can be seen from the symmetric populations of each atom in (b).

Adding an extra atom to the system gives the three-atom model shown in Fig. 5.7, which has 27 coupled energy levels. The dipole–dipole interactions are now dependent on the geometry of the atoms, however to see the effect of blockade it is sufficient to assume $V(R_{12}) = V(R_{13}) = V(R_{23}) = V$. As before the steady-state populations can be found from the density matrix, which reveals population of states $|ggg\rangle$, $|gge\rangle$, $|ggr\rangle$ and $|ger\rangle$ and their respective permutations. This is expected from the analysis of the state for two-atoms; the blockade means only a single atom can contribute to the dark state, whilst the remaining atoms resonantly scatter on the two-level probe transition.

As well as considering the effect of blockade on resonance, it is also necessary to consider the shape of the spectrum. Figure 5.8 shows the real and imaginary parts of the complex susceptibility as a function of probe laser detuning for 1, 2 and 3 atoms

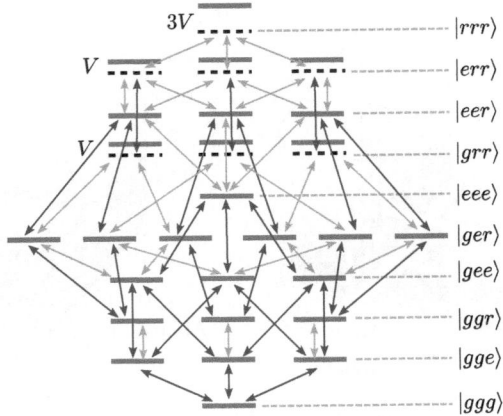

Fig. 5.7 Three-atom model assuming an equilateral geometry where the dipole interaction $V(R_{12}) = V(R_{13}) = V(R_{23}) = V$

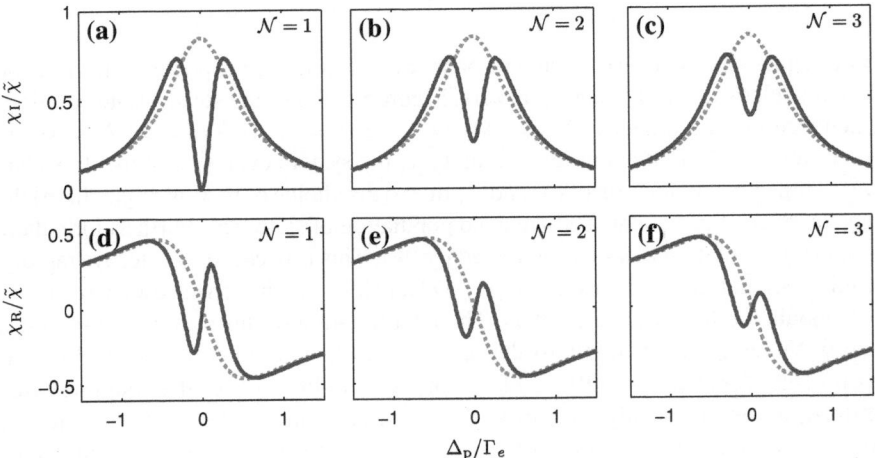

Fig. 5.8 Resonant susceptibility for $\Omega_p = \Gamma_e/4$, $\Omega_c = \Gamma_e/2$ and $V = 2\Gamma_e$, with the probe-only susceptibility plotted as a *dashed line*. χ is scaled relative to the weak, probe-only resonant susceptibility $\tilde{\chi} = 2\rho d_{eg}^2/\epsilon_0 \hbar \Gamma_e$

compared to the probe-only susceptibility. This reveals a suppression in the resonant transmission due to the blockade, associated with a concomitant modification the dispersive lineshape in χ_R. An interesting feature of the susceptibility is that there is no shift or broadening of the two photon resonance, which may be expected as the interactions cause a detuning of the Rydberg pair states. This is because the shifted pair states, known as *anti-blockade* states, become resonant at $\Delta_p = V(R)/2$ which is in the wings of the two-photon resonance when the blockade condition is met.

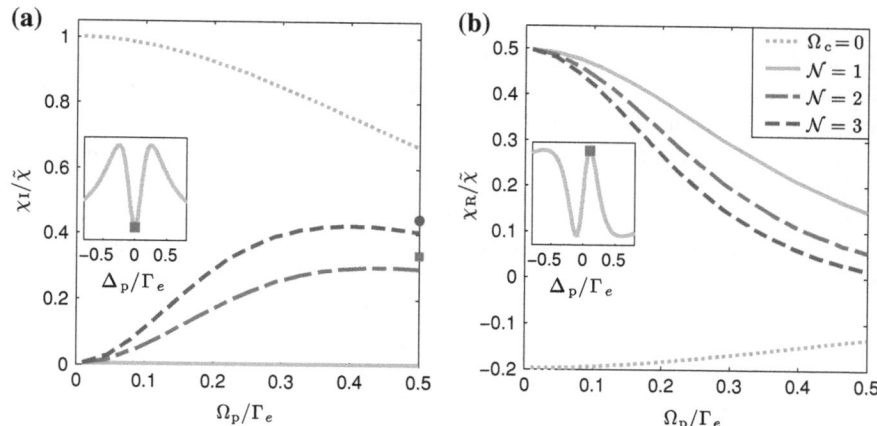

Fig. 5.9 Cooperative non-linearity. **a** Resonant susceptibility calculated for $\Omega_c = \Gamma_e/2$, showing saturation above $\Omega_p = 0.3\,\Gamma_e$. *Single points* represent the empirical scaling χ_{emp} of Eq. 5.22; **b** dispersive non-linearity on the peak of the refractive index feature. χ is scaled relative to the probe-only susceptibility, $\tilde{\chi}$

5.4.3 Cooperative Optical Non-Linearity

The effect of dipole blockade is therefore to change the optical properties of the medium from being perfectly transparent on the EIT resonance to having all but one of the atoms resonantly coupled to the probe beam, suppressing the transmission as the probe power is increased from the weak probe regime. Figure 5.9 shows the susceptibility as a function of Ω_p for a blockade sphere of 1–3 atoms calculated both on resonance and for the dispersive feature. From the resonant susceptibility plotted in (a), it is clear to see the resulting optical non-linearity in the system, which begins to saturate around $\Omega_p = 0.3\,\Gamma_e$ as the two level transition becomes power broadened. The important feature is that the optical non-linearity now depends not only on the probe electric field, but also the number of atoms per blockade sphere. This makes it a cooperative effect, where the optical response of a single atom depends on the surrounding atoms, resulting in the single atom susceptibility $\chi \propto \mathcal{N}_b$, where $\mathcal{N}_b$ is the number of atoms per blockade sphere. Observation of this non-linear density scaling is important as it is this that makes it different to an ordinary non-linear medium.

Previously, a cooperative non-linear effect has only been observed in an up-conversion process [40] requiring large laser intensities to overcome the relaxation mechanisms in the system, making it unsuitable for quantum information processing. The Rydberg states however allow significant tunability of interaction strength through choice of n, enabling the mechanism to be used for very weak probe powers.

Calculating the optical response for $\mathcal{N} > 3$ is challenging as the Hilbert space in the coupled basis scales as $3^{\mathcal{N}}$, rapidly becoming intractable. One method that can be used to reduce the basis states in a many-body system is to use a mean-field theory, where each atom is modified by the mean interaction of all surrounding atoms. This approach has been used successfully to reproduce excitation suppression due to blockade in cold gases [22, 41, 42]. For the EIT system however, it cannot reproduce the spectrum as the mean interaction detunes the $|r\rangle$ states of the atoms, causing the lineshape to shift and broaden [43]—an effect not observed in the full many-body model.

In the limit that the probe is strong enough to saturate the non-linearity, an empirical scaling can be introduced to estimate the susceptibility of a blockade with $\mathcal{N}_b$ atoms,

$$\chi_{\text{emp}} = \frac{1}{\mathcal{N}_b}\chi_1 + \frac{(\mathcal{N}_b - 1)}{\mathcal{N}_b}\chi_1^{(\Omega_c=0)}, \tag{5.22}$$

where χ_1 denotes the susceptibility of a single non-interacting atom. This scaling arises from the fact only a single atom can contribute to the dark state, whilst $\mathcal{N}_b - 1$ atoms resonantly absorb light like an effective two-level atom. This scaling is plotted on Fig. 5.9a as solid markers for $\Omega_p = 0.5\,\Gamma_e$, showing approximate agreement with the exact calculation which allows estimation of the maximum suppression in the system. Thus for large $\mathcal{N}_b$ it should be possible to suppress the transmission to the probe-only value.

Recently a Monte-Carlo method has been developed by Ates et al. to calculate the steady-state density matrix for very large atom numbers [44]. In this work, the authors show that the total susceptibility on resonance scales quadratically with atomic density. This non-linear density scaling represents clear evidence of a cooperative effect, rather than simply a collective scaling.

5.5 Summary

Cooperative phenomena arise when the interaction between the dipole of each atom and the electric field of the surrounding atoms dominates, requiring $kR < 1$. For the low n Rydberg states, this is manifested as superradiance, with the decay reaching a maximum rate of $\mathcal{N}^2\,\Gamma_r$ when half the atoms have decayed. At high n, the energy shift dominates over enhanced dephasing, leading to the dipole blockade which prevents creation of more than a single Rydberg excitation within a radius $R_b = \sqrt[6]{C_6/\Omega}$. This deterministic process is important for applications in quantum information processing, as it allows atomic quantum gates to be implemented. For the case of EIT, blockade modifies the dark state, suppressing the transmission on resonance. This creates a cooperative optical non-linearity in which the optical response of a single atom is modified by the neighbouring atoms, resulting in a non-linear density scaling in the single atom response. In the experiments presented in Chap. 7 the non-linear density scaling and optical non-linearity will be tested for these signatures of cooperativity.

References

1. L. Mandel, E. Wolf, *Optical Coherence and Quantum Optics* (CUP, Cambridge, 2008)
2. J.D. Jackson, *Classical Electrodynamics*, 3rd edn. (Wiley, New York, 1999)
3. M. Born, E. Wolf, *Principles of Optics* (CUP, Cambridge, 1999)
4. R. Loudon, *The Quantum Theory of Light*, 3rd edn. (OUP, Oxford, 2008)
5. R.H. Dicke, Coherence in spontaneous radiation processes. Phys. Rev. **93**(1), 99 (1954)
6. I.E. Protsenko, Superradiance of trapped atoms. J. Russ. Laser Res. **27**(5), 414 (2006)
7. R.G. DeVoe, R.G. Brewer, Observation of superradiant and subradiant spontaneous emission of two trapped ions. Phys. Rev. Lett. **76**(12), 2049 (1996)
8. R.H. Lehmberg, Radiation from an N-atom system. I. General formalism. Phys. Rev. A **2**(3), 883 (1970)
9. N.E. Rehler, J.H. Eberly, Superradiance. Phys. Rev. A **3**(5), 1735 (1971)
10. M. Gross, S. Haroche, Superradiance: an essay on the theory of collective spontaneous emission. Phys. Rep. **93**(5), 301 (1982)
11. J.O. Day, E. Brekke, T.G. Walker, Dynamics of low-density ultracold Rydberg gases. Phys. Rev. A **77**(5), 052712 (2008)
12. N. Skribanowitz, I.P. Herman, J.C. MacGillivray, M.S. Feld, Observation of Dicke superradiance in optically pumped HF gas. Phys. Rev. Lett. **30**(8), 309 (1973)
13. M. Gross, C. Fabre, P. Pillet, S. Haroche, Observation of near-infrared Dicke superradiance on cascading transitions in atomic sodium. Phys. Rev. Lett. **36**(17), 1035 (1976)
14. F. Gounand, M. Hugon, P.R. Fournier, J. Berlande, Superradiant cascading effects in rubidium Rydberg levels. J. Phys. B **12**(4), 547 (1979)
15. T. Wang, S.F. Yelin, R. Côté, E.E. Eyler, S.M. Farooqi, P.L. Gould, M. Koštrun, D. Tong, D. Vrinceanu, Superradiance in ultracold Rydberg gases. Phys. Rev. A **75**(3), 033802 (2007)
16. M.D. Lukin, M. Fleischhauer, R. Cote, L.M. Duan, D. Jaksch, J.I. Cirac, P. Zoller, Dipole blockade and quantum information processing in mesoscopic atomic ensembles. Phys. Rev. Lett. **87**(3), 037901 (2001)
17. D. Jaksch, J.I. Cirac, P. Zoller, Fast quantum gates for neutral atoms. Phys. Rev. Lett. **85**(10), 2208 (2000)
18. D. Møller, L.B. Madsen, K. Mølmer, Quantum gates and multiparticle entanglement by Rydberg excitation blockade and adiabatic passage. Phys. Rev. Lett. **100**(17), 170504 (2008)
19. M. Müller, I. Lesanovsky, H. Weimer, H.P. Büchler, P. Zoller, Mesoscopic Rydberg gate based on electromagnetically induced transparency. Phys. Rev. Lett. **102**(17), 170502 (2009)
20. I.E. Mazets, G. Kurizki, Multiatom cooperative emission following single-photon absorption: Dicke-state dynamics. J. Phys. B **40**(6), F105 (2007)
21. L.H. Pedersen, K. Mølmer, Few qubit atom-light interfaces with collective encoding. Phys. Rev. A **79**(1), 012320 (2009)
22. D. Tong, S.M. Farooqi, J. Stanojevic, S. Krishnan, Y.P. Zhang, R. Côté, E.E. Eyler, P.L. Gould, Local blockade of Rydberg excitation in an ultracold gas. Phys. Rev. Lett. **93**(6), 063001 (2004)
23. K. Singer, M. Reetz-Lamour, T. Amthor, L.G. Marcassa, M. Weidemüller, Suppression of excitation and spectral broadening induced by interactions in a cold gas of Rydberg atoms. Phys. Rev. Lett. **93**(16), 163001 (2004)
24. K. Afrousheh, P. Bohlouli-Zanjani, D. Vagale, A. Mugford, M. Fedorov, J.D.D. Martin, Spectroscopic observation of resonant electric dipole–dipole interactions between cold Rydberg atoms. Phys. Rev. Lett. **93**(23), 233001 (2004)
25. T. Cubel Liebisch, A. Reinhard, P.R. Berman, G. Raithel, Atom counting statistics in ensembles of interacting Rydberg atoms. Phys. Rev. Lett. **95**(25), 253002 (2005)
26. T. Vogt, M. Viteau, J. Zhao, A. Chotia, D. Comparat, P. Pillet, Dipole blockade at Förster resonances in high resolution laser excitation of Rydberg states of cesium atoms. Phys. Rev. Lett. **97**(8), 083003 (2006)
27. T. Vogt, M. Viteau, A. Chotia, J. Zhao, D. Comparat, P. Pillet, Electric-field induced dipole blockade with Rydberg atoms. Phys. Rev. Lett. **99**(7), 073002 (2007)

28. C.S.E. van Ditzhuijzen, A.F. Koenderink, J.V. Hernández, F. Robicheaux, L.D. Noordam, H.B. van Linden van den Heuvell, Spatially resolved observation of dipole–dipole interaction between Rydberg atoms. Phys. Rev. Lett. **100**(24), 243201 (2008)

29. R. Heidemann, U. Raitzsch, V. Bendkowsky, B. Butscher, R. Low, T. Pfau, Rydberg excitation of Bose–Einstein condensates. Phys. Rev. Lett. **100**(3), 033601 (2008)

30. R. Heidemann, U. Raitzsch, V. Bendkowsky, B. Butscher, R. Low, L. Santos, T. Pfau, Evidence for coherent collective Rydberg excitation in the strong blockade regime. Phys. Rev. Lett. **99**(16), 163601 (2007)

31. U. Raitzsch, V. Bendkowsky, R. Heidemann, B. Butscher, R. Low, T. Pfau, Echo experiments in a strongly interacting Rydberg gas. Phys. Rev. Lett. **100**(1), 013002 (2008)

32. M. Reetz-Lamour, T. Amthor, J. Deiglmayr, M. Weidemüller, Rabi oscillations and excitation trapping in the coherent excitation of a mesoscopic frozen Rydberg gas. Phys. Rev. Lett. **100**(25), 253001 (2008)

33. E. Urban, T.A. Johnson, T. Henage, L. Isenhower, D.D. Yavuz, T.G. Walker, M. Saffman, Observation of Rydberg blockade between two atoms. Nat. Phys. **5**, 110 (2009)

34. A. Gaëtan, Y. Miroshnychenko, T. Wilk, A. Chotia, M. Viteau, D. Comparat, P. Pillet, A. Browaeys, P. Grangier, Observation of collective excitation of two individual atoms in the Rydberg blockade regime. Nat. Phys. **5**, 115 (2009)

35. T. Wilk, A. Gaëtan, C. Evellin, J. Wolters, Y. Miroshnychenko, P. Grangier, A. Browaeys, Entanglement of two individual neutral atoms using Rydberg blockade. Phys. Rev. Lett. **104**(1), 010502 (2010)

36. L. Isenhower, E. Urban, X.L. Zhang, A.T. Gill, T. Henage, T.A. Johnson, T.G. Walker, M. Saffman, Demonstration of a neutral atom controlled-NOT quantum gate. Phys. Rev. Lett. **104**(1), 010503 (2010)

37. I. Friedler, D. Petrosyan, M. Fleischhauer, G. Kurizki, Long-range interactions and entanglement of slow single-photon pulses. Phys. Rev. A **72**, 043803 (2005)

38. E. Shahmoon, G. Kurizki, M. Fleischhauer, D. Petrosyan, Strongly interacting photons in hollow-core waveguides. Phys. Rev. A **83**, 033806 (2011)

39. G.B. Arfken, *Mathematical Methods for Physicists*, 3rd edn (Academic Press, New York, 1985)

40. M.P. Hehlen, H.U. Güdel, Q. Shu, J. Rai, S. Rai, S.C. Rand, Cooperative bistability in dense, excited atomic systems. Phys. Rev. Lett. **73**(8), 1103 (1994)

41. H. Weimer, R. Löw, T. Pfau, H.P. Büchler, Quantum critical behavior in strongly interacting Rydberg gases. Phys. Rev. Lett. **101**(25), 250601 (2008)

42. A. Chotia, M. Viteau, T. Vogt, D. Comparat, P. Pillet, Kinetic Monte Carlo modelling of dipole blockade in Rydberg excitation experiment. New J. Phys. **10**, 045031 (2008)

43. H. Schempp, G. Günter, C.S. Hofmann, C. Giese, S.D. Saliba, B.D. DePaola, T. Amthor, M. Weidemüller, S. Sevinçli, T. Pohl, Coherent population trapping with controlled interparticle interactions. Phys. Rev. Lett. **104**(17), 173602 (2010)

44. C. Ates, S. Sevinçli, T. Pohl, Electromagnetically induced transparency in strongly interacting Rydberg gases. Phys. Rev. A **83**(4), 041802 (2011)

Part II
Observations of Cooperativity

Chapter 6
Experiment Setup

There are two requirements for experimental observation of cooperative optical effects in Rydberg EIT due to dipole–dipole interactions. Firstly, a high atomic density is required to ensure a large optical depth on the probe transition whilst meeting the condition $k'R < 1$, where k' is the wave-vector for the Rydberg dipole; and secondly, the dipole–dipole interactions must dominate over any other dephasing mechanisms in the system such as collisional broadening or the Doppler shift. The Doppler effect is important as atoms moving at different velocities observe different laser frequencies. This can shift the blockaded-states back into resonance, leading to a significant reduction of the blockade size for room temperature samples [1].

Both of these requirements can be met using the techniques of laser cooling [2], with which alkali-metal atoms can easily be cooled to temperatures around $100\,\mu K$ with densities in the region of $10^{10}\,cm^{-3}$. At this density, the average interatomic separation $\langle R \rangle \sim \rho^{-1/3} \sim 2\,\mu m$, smaller than the typical blockade radius of $R_b \sim 5\,\mu m$. In this regime the sample can be treated as a *frozen* Rydberg gas [3, 4] where the Rydberg interactions represent the largest energy scale in the system, allowing studies of the excitation dynamics e.g.resonant energy transfer [3, 5–7], mechanical effects of dipole–dipole interactions [8, 9], dipole blockade [10–19] and formation of long-range molecules [20, 21]. These ultra-cold samples are also ideal for precision measurements of quantum defects [22–24] and lifetimes [25, 26] of the Rydberg states.

The experiments presented in this thesis are all performed using the setup shown schematically in Fig. 6.1. Atoms are cooled using three pairs of retro-reflected, counter propagating beams which intersect at the centre of a Kimball Physics spherical octagon vacuum chamber. This chamber was originally designed for a CO_2 lattice experiment [27], and therefore has no field plates for controlling electric fields in the chamber, nor any form of ion detection. The chamber is sealed and pumped down to a pressure below 10^{-10} torr as measured using an ion gauge, with rubidium dispensers mounted inside to provide a source of atoms. The probe beam is aligned through the centre of the atom cloud, monitoring the transmission using a photodiode on the opposite side of the chamber. Atoms are prepared using an optical pumping beam

J. D. Pritchard, *Cooperative Optical Non-Linearity in a Blockaded Rydberg Ensemble*,
Springer Theses, DOI: 10.1007/978-3-642-29712-0_6,
© Springer-Verlag Berlin Heidelberg 2012

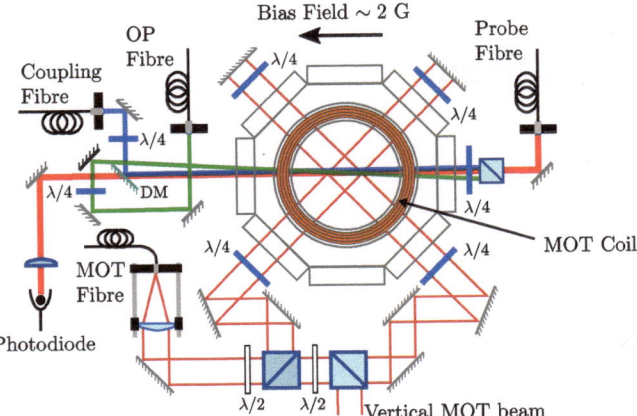

Fig. 6.1 Schematic of experiment setup. Three pairs of orthogonal beams overlap at the centre of a vacuum chamber which is coaxial with a pair of magnetic coils in an anti-Helmholtz configuration to form a magneto-optical trap (MOT) that slows and traps atoms. Following collection of a dense cold atomic sample, the cooling light is extinguished and the atoms are optically pumped, then probed using counter propagating probe and coupling lasers which are overlapped using a dichroic mirror (DM)

which counter-propagates with the probe beam at a shallow angle. The EIT coupling laser is also aligned to counter-propagate with the probe beam. A detailed description of each stage is presented below.

6.1 Laser Cooling

The starting point of any experiment with cold atoms is typically a magneto-optical trap (MOT) [28] both to cool the atoms and provide 3D confinement, resulting in a cold dense gas. Full details can be found in atomic physics textbooks e.g.[29], however a brief description follows. The MOT consists of a pair of magnetic coils in an anti-Helmholtz configuration to create a magnetic quadrupole field. At the origin the field is zero, however the field gradient is linear in all directions. Three orthogonal pairs of counter-propagating circularly polarised laser beams intersect the centre of the coils, aligned so that the vertical beams are coaxial with the coils. The laser light is red-detuned from the atomic transition ($\omega < \omega_0$), such that as atoms move out from the origin the Zeeman-shift due to the magnetic field brings the atoms closer to resonance with the laser field propagating in the opposite direction. This creates a position dependent restoring force which returns the atom to the centre of the beams, trapping the atoms. Cooling occurs due to the Doppler effect, which causes the atom to be shifted closer to resonance with the laser counter-propagating with the direction of motion. This results in a frictional force proportional to atomic velocity that slows the atoms. Combining these mechanisms, atoms can theoretically be cooled down

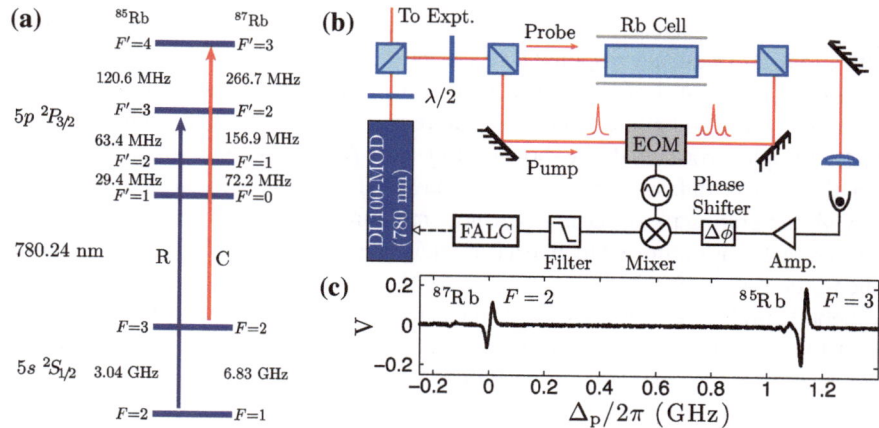

Fig. 6.2 Laser Cooling. **a** Energy levels of rubidium D_2-line [31, 32] showing cooling transition (C) from $5s^2S_{1/2}$ $F = I + 1/2$ to $5p^2P_{3/2}$ $F' = I + 3/2$ and repump transition (R) from $F = I - 1/2$ to $F' = I + 1/2$; **b** schematic of modulation transfer spectroscopy used for locking cooling laser. The probe and pump beams have powers of $500\,\mu$W and $1.2\,$mW respectively; **c** error-signal obtained after demodulation. The large dispersive features correspond to the $F' = I + 1/2$ cooling transition for each isotope

to the Doppler limited temperature of $T_D = \hbar \Gamma_e / 2k_B$, however in practise a MOT can get well below this limit for atoms with hyperfine structure [30].

6.1.1 Cooling Lasers

Rubidium has two natural isotopes, ^{85}Rb and ^{87}Rb, with nuclear spins I of $5/2$ and $3/2$ respectively. The corresponding energy levels for the D_2 line from $5s$ $^2S_{1/2}$ to $5p$ $^2P_{3/2}$ are shown in Fig. 6.2a, with the hyperfine splitting energies taken from D. Steck [31, 32]. Cooling is performed using the closed transition from $F = I + 1/2$ to $F' = I + 3/2$ (C) at 780.24 nm, however some atoms can fall into the $F = I - 1/2$ lower hyperfine ground-state due to an off-resonant excitation of $F' = I + 1/2$. It is therefore necessary to use a repump laser (R) on the transition from $F = I - 1/2$ to $F' = I + 1/2$ to prevent atoms being lost from the cooling cycle.

The cooling light is derived from a Toptica DL-100-MOD diode laser which is stabilised to the closed transition using modulation transfer spectroscopy [33]. The lock setup is shown schematically in Fig. 6.2b, where orthogonally polarised pump and probe beams counter-propagate through an atomic vapour cell. A homebuilt electro-optic modulator (EOM) is used to add side-bands onto the pump laser at a frequency of 9.5 MHz with a modulation index of 0.2. Inside the cell, only atoms with a velocity component along the axis of light propagation with $|kv| < \Gamma_e$ interact with both pump and probe lasers, leading to a four-wave mixing process in which the sidebands are transferred onto the probe laser on the $F = I + 1/2$ to

$F' = I + 3/2$ resonance. The probe beam is detected using a homebuilt photodiode with a 15 MHz frequency response (see appendix A.1). This gives the usual saturation spectroscopy signal, however when demodulated and low-pass filtered at 50 kHz gives a narrow sub-Doppler dispersion feature on a zero-background, removing offset drift. Example error-signals are shown in Fig. 6.2c. A Toptica FALC module is used to lock the laser, with fast-current feedback via a FET on the diode and slow correction through the grating piezo. Using beat-note measurements with two different lasers, the feedback was optimised to give a Lorentzian laser linewidth of $\gamma_p/2\pi = 300$ kHz when averaging over a 20 s period. Light for the repump transition is derived from a homebuilt diode laser, which is stabilised using dichroic-atomic-vapour laser locking (DAVLL) [34]. This setup is described in Sect. 5.3.1 of [27]. Using these techniques both lasers can be locked to either isotope.

6.1.2 MOT

The repump and cooling light are combined on a polarising beam splitter (PBS) with orthogonal polarisations and coupled into a single mode polarisation maintaining fibre which delivers light to the vacuum chamber. The MOT light is then expanded to a $1/e^2$ radius of 9 mm and separated on PBS cubes into three beams that pass orthogonally through the chamber to overlap in the centre, as shown in Fig. 6.1. These are then retro-reflected after the chamber to create counter-propagating beams with orthogonal circular polarisations. Each beam has 8.7 mW of cooling light, however, due to the orthogonal polarisation of the repump light coupled into the fibre, the 6 mW of repump light is distributed unequally among the three beam pairs. A pair of water-cooled coils are mounted coaxially onto the vacuum chamber that create a quadrupole field with a calculated gradient of 0.136 G/cm/A at the centre, which agrees well with the current required for trapping in Sect. 6.2. Additionally, three pairs of rectangular bias coils are arranged around the chamber to cancel the offset magnetic field inside the chamber. Atoms are loaded into the MOT from the background vapour provided from dispensers (SAES Getters) which contain both isotopes in their natural abundance. Once atoms are trapped in the MOT, it is necessary to characterise the atom cloud to find the optimum parameters for cooling.

To determine the temperature and number of atoms in the MOT, a calibrated IR CCD camera (JAI CV-M50) is used to perform fluorescence imaging. This is setup with ×4.6 magnification to give an effective field of view of 5×5 mm, allowing imaging after expansion times of up to 40 ms. The atom number is calculated from summing over the pixel counts using

$$N_{\text{atom}} = \frac{\kappa}{(\Omega/4\pi)\Gamma_{\text{sc}}\tau} \sum_{\text{px}} c_{\text{px}}, \tag{6.1}$$

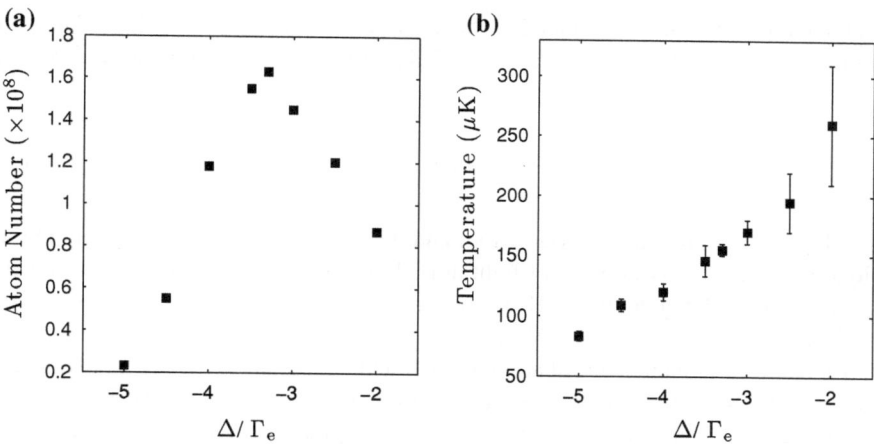

Fig. 6.3 MOT atom number and temperature as a function of detuning for a 5 s load time

where κ is the camera quantum efficiency, Γ_{sc} is the photon scattering rate for each atom, $\Omega/4\pi$ is the collection efficiency and τ is the imaging duration. The quantum efficiency was measured by pulsing a probe beam of known power onto the CCD and fitting the pixel sum as a function of incident photon number to obtain $\kappa = 62$ photons/count. The collection efficiency of the imaging lens is $\Omega/4\pi \simeq r^2/4d^2$, where $r = 13$ mm is the lens radius and $d = 200$ mm the working distance, giving an efficiency of 0.1 %. Finally, the scattering rate is calculated from the excited-state probability multiplied by the decay rate Γ_e, which using the steady-state excited state population (Eq. 4.16a) gives

$$\Gamma_{sc} = \Gamma_e \sigma_{ee}^{ss} = \frac{\Gamma_e}{2} \frac{I/I_{sat}}{1 + I/I_{sat} + (2\,\Delta_p/\Gamma_e)^2}, \tag{6.2}$$

where the relation $\Omega_p = \Gamma_e \sqrt{I/2I_{sat}}$ has been used and I_{sat} is the saturation intensity, defined as [29]

$$I_{sat} \equiv \frac{2\pi\hbar\,\Gamma_e c}{3\lambda^3}. \tag{6.3}$$

Using the decay rate from $5P_{3/2}$ of $\Gamma_e/2\pi = 6.065$ MHz [35] gives $I_{sat} = 1.6$ mW/cm^2. This is only correct for the closed σ^+-transition from $(F = I + 1/2, m_F = F)$ to $(F' = I + 3/2, m'_F = F')$, however in the MOT the atoms are distributed unevenly over a range of m_F states [36]. An effective saturation intensity is used instead, averaging over the transition strengths from all possible m_F levels[1] to give 3.9 mW/cm^2 for ^{85}Rb [31] and 3.6 mW/cm^2 for ^{87}Rb [32].

The temperature of the atoms can be found using time of flight imaging [38], allowing the atoms to expand for a fixed time and fitting a Gaussian profile to the

---

[1] A useful discussion of this is presented in Sect. 4.1 of [37].

cloud to find the radius, σ_r, which is defined as the standard deviation of the Gaussian profile. Extracting the radius for a range of flight times Δt, the temperature T is then determined by fitting to the function

$$\sigma_r(\Delta t)^2 = \sigma_r(0)^2 + \frac{k_B T}{m} \Delta t^2, \tag{6.4}$$

which gives both the initial cloud width and the temperature. To avoid errors in the cloud size due to re-scattering of light at high densities, the atoms are imaged off-resonance by pulsing on the MOT beams for $100\,\mu s$ at a detuning of $\Delta_p = -\Gamma_e$ for $I = 1.5 I_{sat}$ to give $\Gamma_{sc}/2\pi = 1\,\text{MHz}$.

Using the two diagnostics of atom number and temperature, the MOT parameters were optimised to give the greatest number of atoms, and hence optical depth along the probe laser. Data taken for a 5 s load time at a gradient of 13.5 G/cm are shown in Fig. 6.3 as a function of detuning, in units of the decay rate of the $5P_{3/2}$ state Γ_e. The coldest temperatures are obtained at large detuning, giving cooling below the Doppler limit $T_D = 140\,\mu K$, however at the cost of atom number. The MOT detuning was therefore set at $\Delta = -3.5\Gamma$, with a temperature of $150\,\mu K$.

6.1.3 Optical Molasses

Once the quadrupole field of the MOT is turned off the atoms are cooled by the radiation pressure of the MOT beams, known as an optical molasses. In the molasses, temperatures far below the Doppler limit are achieved for multi-level atoms due to the spatially dependent polarisation gradients formed by the interference of the circularly polarised beams at the centre of the trap [39]. The temperature obtained in the molasses is related to the dimensionless light-shift parameter $\Omega^2/|\Delta|\Gamma$ by [36]

$$\frac{k_B T}{\hbar \Gamma} = C_{\sigma^+ \sigma^-} \frac{\Omega^2}{|\Delta|\Gamma} + C_0, \tag{6.5}$$

where Ω is the Rabi frequency calculated using the intensity per beam. The minimum temperature achievable in sub-Doppler cooling is limited by the intensity of the light becoming so weak that the atom does not experience a polarisation gradient. This causes the linear relationship between temperature and light-shift to break down below $\Omega^2/|\Delta|\Gamma \simeq 0.05$, and the temperature approaches that of the MOT.

In order to test the relationship of Eq. 6.5 for ^{87}Rb, the molasses was first optimised by pulsing the quadrupole MOT field on and off and monitoring the cloud expansion on a camera in real time. The 3 bias coils were then set to give a slow, isotropic expansion of the cloud when the coil is off, corresponding to zero magnetic field at the chamber centre. If the field is not cancelled, the Zeeman splitting due to the residual field leads to the atom being optically pumped into a particular m_F state and preferentially absorbing light from one direction, accelerating the atoms out of the

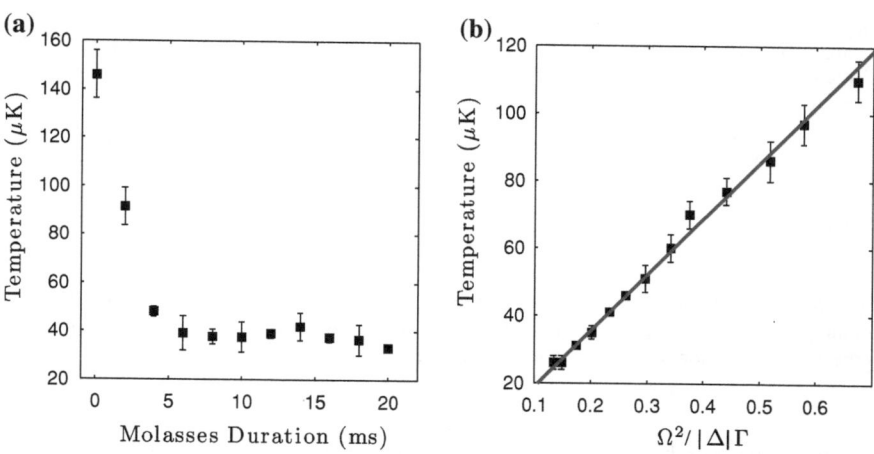

Fig. 6.4 a Temperature versus molasses duration for molasses temperature as a function of dimensionless light shift parameter $\Omega^2/|\Delta|\Gamma$. Fitting gives $C_{\sigma^+\sigma^-} = 0.58 \pm 0.02$

molasses beams. Atoms were then loaded into the MOT for 5 s and the temperature as a function of molasses duration found. The results for I (per beam)$= 3.3\,\mathrm{mW/cm^2}$ and $\Delta = -10\,\Gamma_e$ are presented in Fig. 6.4a, showing that the molasses temperature changes significantly in the first 10 ms then remains approximately constant. Temperature was then measured after a 20 ms molasses duration for a variety of values of the light-shift parameter, varying both I and Δ, with results shown in Fig. 6.4b. Each point represents the mean and standard error of 8 measurements, which were used to perform a χ^2 fit to give $C_{\sigma^+\sigma^-} = 0.58 \pm 0.02$. This was calculated using $I_{sat} = 3.2\,\mathrm{mW/cm^2}$ to allow direct comparison with the results of Wallace et al. [40] who measured $C_{\sigma^+\sigma^-} = 0.52 \pm 0.03$.

Subsequent improvement of the cancellation fields and beam-balance has further reduced the temperature to 20 μK for a 1 s MOT load and 10 ms molasses duration, resulting in an atom cloud with a Gaussian width of σ_r of 0.5–0.7 mm.

6.2 Optical Pumping

As mentioned above, in the MOT and molasses atoms are distributed over a range of m_F states due to their interaction with the light-field. For the EIT experiments however, the atoms need to be prepared into the $5s\,^2S_{1/2}$ ($F = I + 1/2, m_F = I + 1/2$) ($|g\rangle$) stretched state to give the strongest coupling to the $5p\,^2P_{3/2}$ ($F = I + 3/2, m_F = I + 3/2$) state on the closed σ^+-transition. Preparing the sample in a single state has two other advantages—firstly, it simplifies numerical modelling of the system as there is only a single excitation pathway, and secondly, it prevents excitation of Rydberg pair-states with different m_j values, which experience weak or even zero interaction strengths [41, 42].

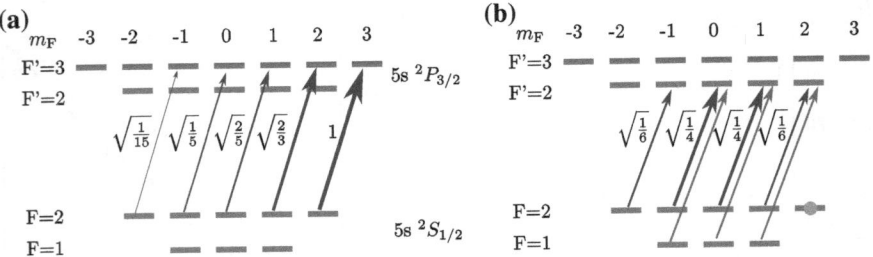

Fig. 6.5 Optical pumping schemes for ^{87}Rb to prepare atoms in ($F = 2, m_F = 2$) **a** bright-state pumping **b** dark state pumping. The transition-strengths for each transition are rescaled with respect to the closed transition from ($F = 2, m_F = 2$) to ($F' = 3, m'_F = 3$).

Two of the possible optical pumping schemes that can be used to achieve this are illustrated in Fig. 6.5 for ^{87}Rb. The simplest is bright-state pumping (a) using the σ^+ cooling transition to pump population across to the stretched state. This requires no additional laser frequencies, however once atoms are pumped into ($F = 2, m_F = 2$) they continue to scatter light from the pumping laser, leading to heating. Another option is dark state pumping (b) using σ^+ light resonant with $F = 1$ to $F' = 2$, which makes $m_F = 2$ a dark state, allowing population to collect in this state without further scattering. As atoms can decay from $F' = 2$ to the lower hyperfine ground-state it is also necessary to use repump light with the same polarisation as the pumping light. Dark-state pumping is therefore a much better method for preparing the atomic sample, however it is necessary to obtain light resonant on the $F = I+1/2$ to $F' = F$ transition.

The frequency of the cooling laser is controlled with a 200 MHz AOM in a double-pass configuration which is set to lock 440 MHz off-resonance. Additional AOMs are then used to control the frequency and intensity of the probe and MOT light independently. As shown in Fig. 6.2a, the detuning required for the dark state pumping in ^{87}Rb is 266.7 MHz which can be achieved using another double-pass AOM at 86.65 MHz, whilst for ^{85}Rb the energy difference is 120.6 MHz, requiring a double pass at 159.7 MHz. The transition in ^{87}Rb is more convenient, so the optical pumping is setup for this isotope.

To measure the efficiency of the optical pumping, the quadrupole coils were used to create a magnetic trap which has a force along z given by [29]

$$F = -g_F \mu_B m_F \frac{\mathrm{d}|\boldsymbol{B}|}{\mathrm{d}z}, \tag{6.6}$$

where g_F is the Landé g-factor [43] and μ_B is the Bohr magneton. For atoms with $m_F g_F > 0$, known as weak-field seeking states, the atoms can be trapped when the force is larger than gravity (as first demonstrated for a cooled Na beam [44]). For atoms in the ($F = 2, m_F = 2$) state $m_F g_F = 1$, requiring a gradient of 15 G/cm to trap them. The other weak-field seeking states are ($F = 2, m_F = 1$) and ($F = 1, m_F = -1$) which both have $m_F g_F = 1/2$, corresponding to a gradient of

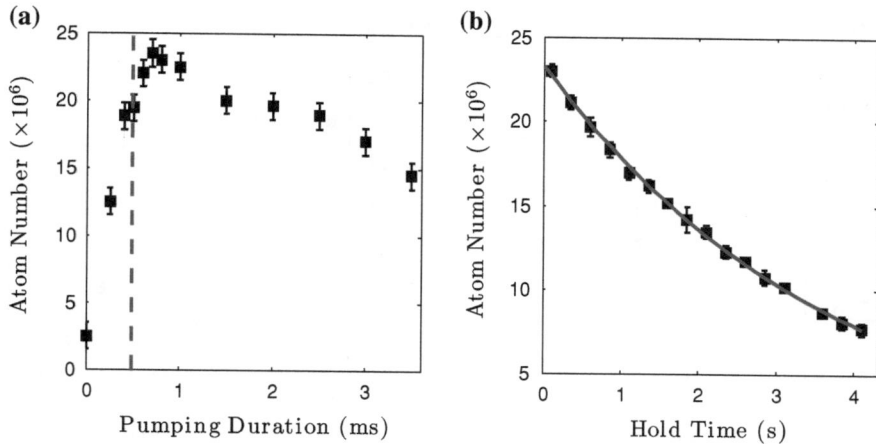

Fig. 6.6 Optical pumping into quadrupole trap. **a** Atom number versus optical pumping duration peaks close to the time required for each atom to scatter an average of 10 photons, indicated by *red dashed line*; **b** atom number versus hold time gives a 1/e lifetime of $\tau = 3.8 \pm 0.3$ s, limited by background collisions

30 G/cm. The maximum gradient for the experiment MOT coils is 20 G/cm, ensuring only the desired state can be trapped.

Light resonant with $F = 2$ to $F' = 2$ and repump light are combined with the same linear polarisation on a 50:50 beam splitter and coupled into a polarisation-maintaining fibre. The fibre output is collimated to a $1/e^2$ beam waist of 1.7 mm to ensure the MOT is approximately uniformly illuminated, and circularly polarised to drive σ^+-transitions. The beam is then aligned into the chamber coaxial to one of the pairs of cancellation coils which provides a magnetic field along the beam axis to define the quantisation axis for the atoms. Atoms are loaded into the MOT for 3 s at a gradient of 20 G/cm, which is then turned off for a 10 ms molasses with a peak atom number of 33×10^6 at 25 μK. Light is extinguished from the chamber for a period of 1 ms to allow the bias-coil to turn on, after which time the optical pumping pulse is applied. The quadrupole field is then turned on and the atoms are then held in the trap for at least 100 ms to let the un-pumped atoms fall away, and imaged to determine the atom number. The bias field, pumping duration and beam powers are then optimised by maximising the atom number after a 100 ms trap time.

Figure 6.6a shows atom number as a function of pumping duration with 400 nW of optical pumping light and 80 μW for the repump transition, with a bias field of 2 G. For short pumping times the number of atoms in the trap is significantly enhanced, obtaining approximately eight times more after 1 ms compared to the unpumped case. This agrees well with the enhancement expected relative to an isotropic distribution of atoms across the ground-state hyperfine levels, for which only 1/8th of the population is expected in the stretched state. Longer pumping times results in a gradual loss due to atoms in the dark state being able to scatter off-resonantly with the $F = 2$ to $F' = 3$ state and decay to un-trapped states. Using the coupling strengths from

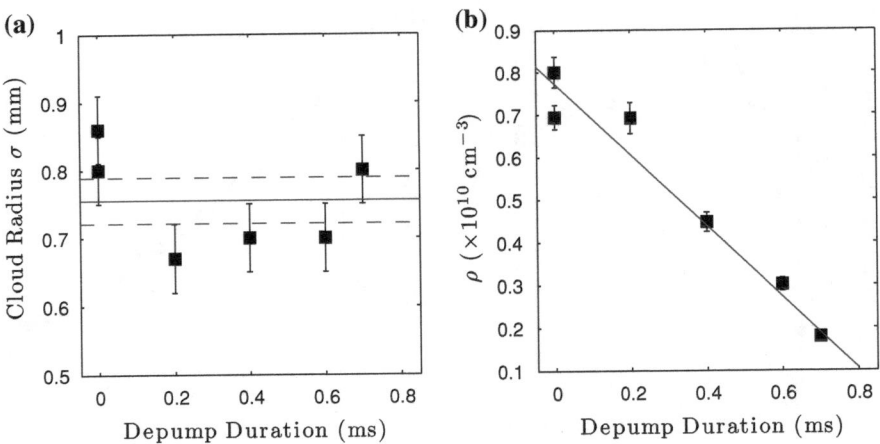

Fig. 6.7 **a** Cloud radius as a function of depump time with mean and standard error plotted in *red*, showing the cloud radius is approximately constant; **b** transmission data shows the density variation is linear with depump duration

Fig. 6.5, the average transition-strength is $\sqrt{1/5}$ which gives a mean scattering rate of 20 kHz for these parameters. Assuming an average of 10 photons is required to pump the atoms across, this should give a peak number around 0.5 ms, consistent with the observed pumping rate.

Fixing the pumping duration as 1 ms, the atom number is relatively insensitive to the bias field for fields above 1 G, similarly for the repump power above 50 μW which is sufficient to repump all of the atoms out of $F = 1$. Additional parameters to improve the atom number are the wave-plate angle and beam alignment to improve the matching to the bias-field. Ideally a retro-reflected optical pumping beam should be used to prevent any heating of the atoms, however this was not possible in the experiment setup. Having optimised the parameters, the atom number is measured as a function of hold time in the trap, Fig. 6.6b. Fitting the data gives a 1/e lifetime of 3.8 ± 0.3 s, which is limited by collisions with background Rb atoms. Extrapolating the lifetime fit to zero hold time, the peak atom number in the trap is 23×10^6, corresponding to 70 % pumping efficiency. This could be partly limited by loss of atoms from the trapping volume during the time delay in which the bias field is switched.

6.2.1 Ground-State Density

Atomic density is an important parameter to vary in the EIT experiments, as it governs the number of atoms per blockade sphere in the sample. Ballistic expansion gives a simple method of varying the ground-state density, however it also changes the length of the sample and allows the atoms to drop under gravity. These effects makes analysis of the transmission variation due to the changing density rather than

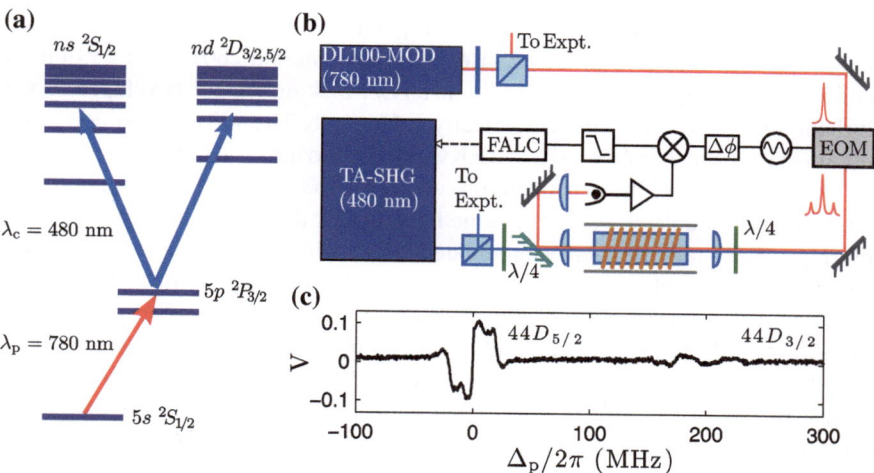

Fig. 6.8 **a** Two-photon excitation scheme with a probe laser at 780 nm on the cooling transition and a coupling laser at 480 nm to the Rydberg states; **b** EIT lock setup using the same EOM as for the modulation transfer lock described above; **c** example error signal

changing optical path length ambiguous. Instead, the repump laser is used to depump atoms into the lower hyperfine ground state by turning it off before the end of the optical pumping pulse. This allows the fraction of atoms in ($F = 2$, $m_F = 2$) to be changed in a controlled way whilst keeping the cloud size fixed.

Figure 6.7a shows the variation of cloud radius as a function of the depump time, which is the time difference between the repump turning off and the end of the optical pumping pulse. The cloud is imaged without any repump light to ensure only atoms in $F = 2$ contribute to the image, which shows the cloud size remains approximately constant during the depump process. Resonant transmission data were recorded simultaneously using the probe beam, from which the average density ρ is extracted (see below), which is plotted in (b). This shows that the density variation is linear with the depump duration.

6.3 Rydberg Excitation

A key component of experiments on Rydberg states is the ability to excite atoms coherently to the Rydberg state. Using a laser at 297 nm atoms can be excited directly from the $5S_{1/2}$ ground state to the np states, however this has two significant disadvantages. The first is that the transition strengths are very low, requiring a high intensity laser; and second the np states do not have isotropic interactions, and for some states have angles with zero interaction [45]. A more convenient scheme is to perform a two-photon excitation via $5P_{3/2}$ shown in Fig. 6.8a, where the first photon is at the laser cooling frequency (780 nm) and a second photon around 480 nm to either ns or nd states, allowing a choice of repulsive or attractive van der Waals interactions (as discussed in Sect. 3.2).

The upper transition is provided by a Toptica TA-SHG laser, which frequency doubles an amplified 960 nm diode laser to provide around 280 mW at 480 nm. To stabilise the laser to the Rydberg transition, an EIT locking scheme developed here in Durham [46] is used, shown schematically in Fig. 6.8b. The 480 nm laser acts as the coupling laser, which is focused into a Rb vapour cell to maximise the Rabi frequency and drives $\sigma^\pm$-transitions dependent on whether an ns or nd state is required. The probe laser is modulated using the same EOM as for the modulation transfer lock, splitting off a 5 μW frequency modulated beam which drives the σ^+-transition. The probe laser counter-propagates with the coupling laser to minimise the Doppler mismatch which creates a frequency shift equal to $(k_c - k_p)v$, where $k_{p,c}$ are the wave-vectors of the probe and coupling lasers, respectively, and v is the velocity along the beam axis. A dichroic mirror picks off the probe beam which is detected using a Hamamatsu C5460 APD module with 20 MHz bandwidth. This signal is amplified and demodulated, finally using a low-pass filter to obtain the error signal. The lock signal is generated from the beat signal between the sidebands and the probe beam that acts as an optical heterodyne resulting in a line-shape similar to Pound-Drever-Hall stabilisation to a cavity [47]. The laser is locked to this signal using a Toptica FALC module to provide fast current modulation to the 960 nm diode and slower correction using the grating piezo.

An example error signal for the $44D_{5/2}$ is shown in Fig. 6.8c, obtained using a 15 mW coupling beam, which shows the technique provides dispersive feature which is much narrower than the 300 MHz Doppler width of the probe transition. The vapour cell is wound inside a solenoid and mounted in a mu-metal shield, which allows a Zeeman shift to be applied to lock on the wings of the EIT resonance if required. One significant advantage of this locking technique is that the EIT signal is generated from a two-photon resonance, meaning that the coupling laser is locked relative to the probe laser. This provides common-mode noise rejection which correlates the frequency fluctuations of the two lasers. As will be shown later, the resulting two-photon Lorentzian linewidth is measured from the cold atom EIT to be $\gamma_{\rm rel}/2\pi = 100$ kHz, a third of the linewidth of the probe laser.

This laser is then coupled into a Schäfter-Kirchhoff (S+K) single-mode polarisation maintaing fibre, providing up to 100 mW at the vacuum chamber. The output coupler has an adjustable collimator (S+K 60FC-4-M12) which allows the beam waist at the centre of the chamber to be varied from a collimated $1/e^2$ radius of 0.8 mm to a strongly focus waist around 70 μm to give a large intensity for coupling to high n-states.

6.4 EIT Experiments

Having obtained a cold and optically pumped atomic ensemble, the optical pumping light is extinguished and the atoms allowed to expand freely for 1 ms. The delay ensures the optical pumping AOM is turned off before the strong 480 nm coupling laser is turned on, preventing any unwanted Rydberg excitation. As there is no AOM

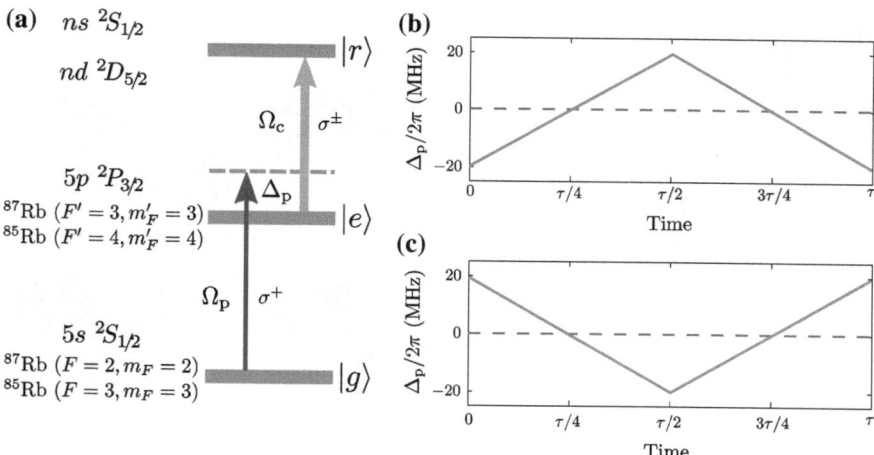

Fig. 6.9 EIT spectroscopy. **a** Level scheme showing the $|g\rangle$, $|e\rangle$ and $|r\rangle$ levels probed in the experiment. **b** Negative-positive-negative (NPN) and **c** positive-negative-positive (PNP) frequency ramps used to give a double-scan across resonance

on the coupling laser, this is shuttered using a homebuilt mechanical shutter [48] which has a $100\,\mu s$ switching time but around $250\,\mu s$ jitter. Once the coupling laser is on, the EIT spectroscopy is performed using the two-photon scheme shown schematically in Fig. 6.9a, with the probe laser driving the closed σ^+-transition from $(F = I + 1/2, m_F = F)$ to $(F' = I + 3/2, m'_F = F')$. To obtain the EIT spectra, the probe laser is scanned ± 20 MHz across the transition in a time τ using the frequency ramps shown in Fig. 6.9b and c whilst keeping the coupling laser on resonance. Spectroscopy is typically performed using a negative-positive-negative (NPN) ramp shown in (b), however for the D-states the scan direction plays an important role, and a positive-negative-positive (PNP) ramp (c) is also used. Using this double-scan technique allows the full EIT spectrum to be obtained in a single experiment, whilst giving information about any loss or hysteresis from the first scan. The frequency ramp is controlled using an Agilent 33250A arbitrary function generator to vary the probe AOM frequency smoothly across the transition. Probe transmission is detected using a Hamamatsu C5460 APD module with 20 MHz bandwidth, which is connected to a Tektronix DPO 4034 digital oscilloscope.

The experiment is computer controlled using LabVIEW to output synchronised digital and analog patterns via a DIO-32HS 32-channel digital output card and PCI-6713 8-channel analog out card. This is interfaced with the oscilloscope to allow automated data acquisition, with the probe power being actively stabilised between experiments using an analog input on the probe AOM attenuator to remove long term drifts in intensity.

For a given set of parameters, data are recorded in three stages—firstly probe-only transmission is recorded with no atoms loaded to obtain the background voltage, averaging over at least 10 repeats. The atoms are then loaded and probe-only data

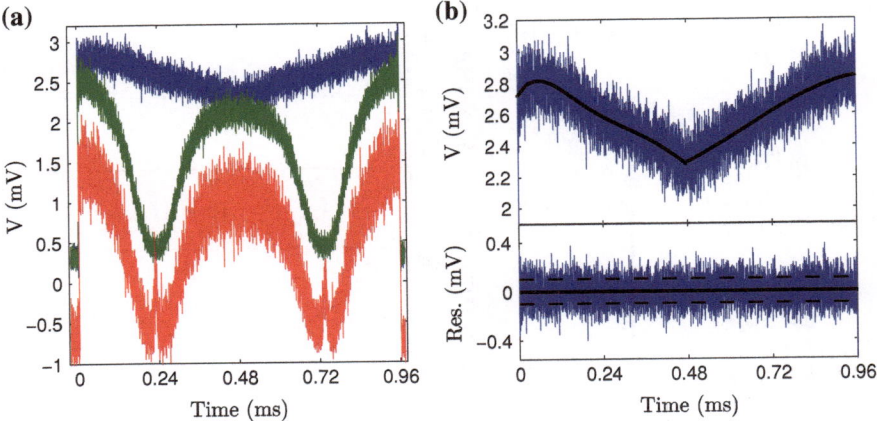

Fig. 6.10 Photodiode data for 5 nW probe beam. **a** Background and absorption traces are 10 shot averages, whilst the EIT (*red*) is a single experiment run vertically offset for clarity by 1 mV; **b** polynomial fit to the background data, with fit residuals

recorded again, also averaging over at least 10 repeats. Finally, data are taken with both probe and coupling lasers, recording EIT spectra as single experiment runs. This is done to prevent averaging out the narrow transparency feature due to fluctuations in absolute frequency of the EIT laser lock, or for parameters close to superradiant behaviour where two repeats give very different results at intermediate probe powers.

6.5 Data Analysis

The photodiode provides a voltage proportional to the power in the probe beam, however the parameter of interest is the transmission. As some of the data are taken at very low probe powers, it is necessary to ensure the errors due to noise in the signal are dealt with correctly to obtain the correct transmission. An example dataset is presented in Fig. 6.10a which shows voltages obtained using a 5 nW probe beam for a $\tau = 0.96$ ms scan time. The background and absorption traces are both taken as 10 shot averages, whilst the EIT data (red) is a single run, offset vertically by 1 mV for clarity. The background probe trace is not flat due to the slight variation in fibre-coupling efficiency as the frequency of the double-pass probe AOM is scanned. Powers are therefore measured at zero detuning, to give the power on the EIT resonance.

Due to the poor signal to noise ratio, directly dividing the absorption voltage data by the background data yields a very noisy transmission signal. Instead, the background data are used to obtain the mean and standard deviation of the intrinsic offset voltage (V_{offs}, σ_{offs}) of the photodiode using the data either side of the probe pulse. The pulse is then divided at $\tau/2 = 0.48$ ms to separate the two scans across

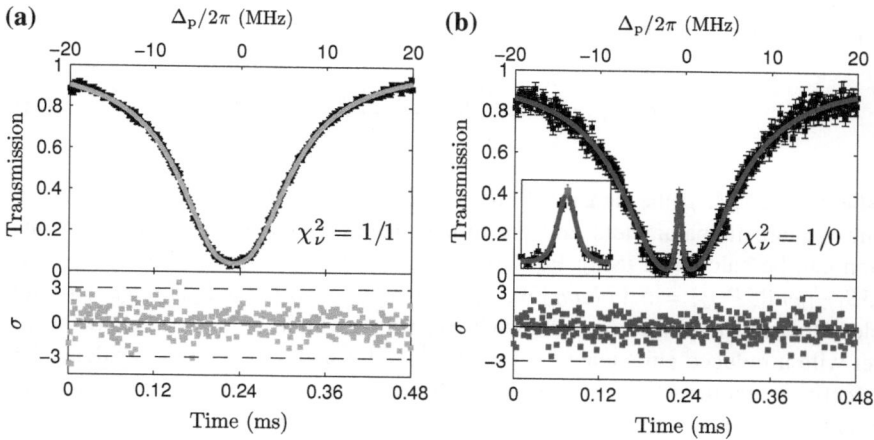

Fig. 6.11 Data Fitting. **a** Fit to probe-only absorption data using Eq. 6.10; **b** EIT transmission data fit using Eq. 6.11. Both fits are well conditioned with structureless residuals

resonance. Each of the datasets is then fitted using a least-squares minimisation to a 10th-order polynomial, shown in Fig. 6.10b. Subtracting the fit from the data, the residuals are used to determine the standard deviation of the fit, σ_{fit}. As the noise during the background comes from the photodiode noise (σ_{offs}) and the laser (σ_{bg}), then providing they are uncorrelated the noise terms will add in quadrature to give the standard-deviation in the fit. The variance in the background is then given by

$$\sigma_{bg}^2 = \sigma_{fit}^2 - \sigma_{offs}^2. \tag{6.7}$$

The absorption signal, V_{abs}, is converted to transmission T using

$$T(V_{atoms}; V_{bg}; V_{offs}) = \left(\frac{V_{atoms} - V_{offs}}{V_{bg} - V_{offs}} \right), \tag{6.8}$$

where V_{bg} is replaced by the best-fit polynomial that was fit to the background. The uncertainty in the transmission, σ_T, can be calculated from [49]

$$\sigma_T^2 = \left\{ T(V_{atoms} + \sigma_{atoms}; V_{bg}; V_{offs}) - T_0 \right\}^2 + \left\{ T(V_{atoms}; V_{bg} + \sigma_{offs}; V_{offs}) - T_0 \right\}^2$$
$$+ \left\{ T(V_{atoms}; V_{bg}; V_{offs} + \sigma_{offs}) - T_0 \right\}^2, \tag{6.9}$$

where $T_0 \equiv T(V_{atoms}; V_{bg}; V_{offs})$ and the uncertainty in the absorption voltage σ_{atoms} is assumed to be the same as the background uncertainty σ_{bg}. Data are then binned in 10-point windows and a weighted average of each bin taken to smooth the data and improve the signal to noise.

Finally, a Levenberg-Marquardt algorithm is used to perform a χ^2-fit of the absorption data to the theoretical two-level transmission profile obtained by combining

Eqs. 4.23a, 4.16b and 4.18 to give

$$T = \exp\left(-\frac{k\alpha d_{eg}^2}{\epsilon_0 \hbar} \frac{\gamma_{eg}}{\Omega_p^2/2 + \gamma_{eg}^2 + \Delta_p^2}\right), \tag{6.10}$$

where $\alpha \equiv \rho\ell$, ρ the density, ℓ the optical path length through the cloud, Δ_p and γ_{eg} are fit parameters and the dipole matrix element for the closed transition can be calculated using Eq. 2.16 combined with the reduced matrix element $\langle 5S_{1/2}||er||5P_{3/2}\rangle = 5.177ea_0$ [50, 51] to give $d_{eg} = \sqrt{1/3} \times 5.177ea_0$. An example fit to the data from above is shown in Fig. 6.11a, which shows a good fit to the data set with a reduced chi-squared value of $\chi_\nu^2 = 1.1$ and structureless residuals. The effective linear density ρ along the probe beam is then calculated from the measured cloud size $\ell = 2\sigma_r$ from the time of flight imaging to give $\rho = \alpha/\ell$, with uncertainty calculated from the standard error. This is how the density measurements are obtained in Fig. 6.7b above.

A similar procedure is applied to the EIT spectra, however as the traces are recorded as a single experiment rather than as an average, it is necessary to increase the size of the errorbar in Eq. 6.9 to give $\sigma_{atoms} = \sqrt{N}\sigma_{bg}$, where N is the number of averages used for the background data. Data are then fit using the weak-probe EIT transmission calculated using the susceptibility from Eq. 4.17

$$T = \exp\left(-\frac{k\alpha d_{eg}^2}{\epsilon_0 \hbar} \mathrm{Im}\left\{\frac{i\gamma_{gr} + (\Delta_p + \Delta_c)}{(\gamma_{eg} - i\Delta_p)(\gamma_{gr} - i(\Delta_p + \Delta_c)) + \Omega_c/4^2}\right\}\right), \tag{6.11}$$

to obtain the parameters γ_{gr}, Ω_c and Δ_c, where α, Δ_p and γ_{eg} are constrained from the absorption fit. An example is shown in Fig. 6.11b, which has larger errorbars than (a) due to the $\sqrt{N}$ factor. As before, the fit shows very good agreement with the data, giving $\chi_\nu^2 = 1.0$ with structureless residuals even around the EIT resonance. The weak-probe formula is therefore an excellent description of the observed spectra, validating the underlying assumption of each atom giving an independent optical response.

A number of values are extracted from these fit parameters. Firstly, for data in the weak probe regime the relative two-photon linewidth can be calculated from $\gamma_{rel} = \gamma_{gr} - \Gamma_r/2$, and similarly the effective probe laser linewidth can be found from $\gamma_p = \gamma_{ge} - \Gamma_e/2$, which typically agrees well with the measured value of 300 kHz. Secondly, the position of the two-photon laser linewidth can be used to determine the transmission on resonance, highlighted as the blue datapoint on the inset of (b).

6.6 Summary

Using the techniques of laser cooling and optical pumping, the atomic sample can be prepared in a well defined initial state from which EIT spectroscopy can be performed. From the transmission spectra, properties such as detunings and linewidths are obtained which give an insight into the cooperative phenomena arising due to the strong dipole–dipole interactions between Rydberg atoms. These cooperative effects will be the subject of the following chapter.

References

1. H. Kübler, J.P. Shaffer, T. Baluktsian, R. Löw, T. Pfau, Coherent excitation of Rydberg atoms in micrometre-sized atomic vapour cells. Nat. Photon. **4**, 112 (2010)
2. C.S. Adams, E. Riis, Laser cooling and trapping of neutral atoms. Prog. Quant. Electron. **21**, 1 (1997)
3. W.R. Anderson, J.R. Veale, T.F. Gallagher, Resonant dipole–dipole energy transfer in a nearly frozen Rydberg gas. Phys. Rev. Lett. **80**, 249 (1998)
4. I. Mourachko, D. Comparat, F. de Tomasi, A. Fioretti, P. Nosbaum, V.M. Akulin, P. Pillet, Many-body effects in a frozen Rydberg gas. Phys. Rev. Lett. **80**, 253 (1998)
5. M. Mudrich, N. Zahzam, T. Vogt, D. Comparat, P. Pillet, Back and forth transfer and coherent coupling in a cold Rydberg dipole gas. Phys. Rev. Lett. **95**, 233002 (2005)
6. S. Westermann, T. Amthor, A.L. de Oliveira, J. Deiglmayr, M. Reetz-Lamour, M. Weidemuller, Dynamics of resonant energy transfer in a cold Rydberg gas. Eur. Phys. J. D **40**, 37 (2006)
7. J.A. Petrus, P. Bohlouli-Zanjani, J.D.D. Martin, ac electric-field-induced resonant energy transfer between cold Rydberg atoms. J. Phys. B **41**, 245001 (2008)
8. T. Amthor, M. Reetz-Lamour, S. Westermann, J. Denskat, M. Weidemüller, Mechanical effect of van der waals interactions observed in real time in an ultracold Rydberg gas. Phys. Rev. Lett. **98**, 023004 (2007)
9. T. Amthor, M. Reetz-Lamour, C. Giese, M. Weidemüller, Modeling many-particle mechanical effects of an interacting Rydberg gas. Phys. Rev. A **76**, 054702 (2007)
10. D. Tong, S.M. Farooqi, J. Stanojevic, S. Krishnan, Y.P. Zhang, R. Côté, E.E. Eyler, P.L. Gould, Local blockade of Rydberg excitation in an ultracold gas. Phys. Rev. Lett. **93**, 063001 (2004)
11. K. Singer, M. Reetz-Lamour, T. Amthor, L.G. Marcassa, M. Weidemüller, Suppression of excitation and spectral broadening induced by interactions in a cold gas of Rydberg atoms. Phys. Rev. Lett. **93**, 163001 (2004)
12. K. Afrousheh, P. Bohlouli-Zanjani, D. Vagale, A. Mugford, M. Fedorov, J.D.D. Martin, Spectroscopic observation of resonant electric dipole–dipole interactions between cold Rydberg atoms. Phys. Rev. Lett. **93**, 233001 (2004)
13. T. Cubel Liebisch, A. Reinhard, P.R. Berman, G. Raithel, Atom counting statistics in ensembles of interacting Rydberg atoms. Phys. Rev. Lett. **95**, 253002 (2005)
14. T. Vogt, M. Viteau, J. Zhao, A. Chotia, D. Comparat, P. Pillet, Dipole blockade at förster resonances in high resolution laser excitation of Rydberg states of cesium atoms. Phys. Rev. Lett. **97**, 083003 (2006)
15. T. Vogt, M. Viteau, A. Chotia, J. Zhao, D. Comparat, P. Pillet, Electric-field induced dipole blockade with Rydberg atoms. Phys. Rev. Lett. **99**, 073002 (2007)
16. C.S.E. van Ditzhuijzen, A.F. Koenderink, J.V. Hernández, F. Robicheaux, L.D. Noordam, H.B. van Linden van den Heuvell, Spatially resolved observation of dipole–dipole interaction between Rydberg atoms. Phys. Rev. Lett. **100**, 243201 (2008)

17. R. Heidemann, U. Raitzsch, V. Bendkowsky, B. Butscher, R. Low, T. Pfau, Rydberg excitation of Bose-Einstein condensates. Phys. Rev. Lett. **100**, 033601 (2008)
18. E. Urban, T.A. Johnson, T. Henage, L. Isenhower, D.D. Yavuz, T.G. Walker, M. Saffman, Observation of Rydberg blockade between two atoms. Nat. Phys. **5**, 110 (2009)
19. A. Gaëtan, Y. Miroshnychenko, T. Wilk, A. Chotia, M. Viteau, D. Comparat, P. Pillet, A. Browaeys, P. Grangier, Observation of collective excitation of two individual atoms in the Rydberg blockade regime. Nat. Phys. **5**, 115 (2009)
20. V. Bendkowsky, B. Butscher, J. Nipper, J.P. Shaffer, R. Löw, T. Pfau, Observation of ultralong-range Rydberg molecules. Nature **458**, 1005 (2009)
21. K.R. Overstreet, A. Schwettmann, J. Tallant, D. Booth, J.P. Shaffer, Observation of electric-field-induced Cs Rydberg atom macrodimers. Nat. Phys. **5**, 581 (2009)
22. W. Li, I. Mourachko, M.W. Noel, T.F. Gallagher, Millimeter-wave spectroscopy of cold Rb Rydberg atoms in a magneto-optical trap: Quantum defects of the ns, np and nd series. Phys. Rev. A **67**, 052502 (2003)
23. J. Han, Y. Jamil, D.V.L. Norum, P.J. Tanner, T.F. Gallagher, Rb nf quantum defects from millimeter-wave spectroscopy of cold ^{85}Rb Rydberg atoms. Phys. Rev. A **74**, 054502 (2006)
24. B. Sanguinetti, H.O. Majeed, M.L. Jones, B.T.H. Varcoe, Precision measurements of quantum defects in the $n\mathrm{P}_{3/2}$ Rydberg states of ^{85}Rb. J. Phys. B **42**, 165004 (2009)
25. D.B. Tretyakov, I.I. Beterov, V.M. Entin, I.I. Ryabtsev, P.L. Chapovsky, Investigation of cold Rb Rydberg atoms in a magneto-optical trap. J. Expt. Th. Phys. **108**, 374 (2008)
26. D.B. Branden, T. Juhasz, T. Mahlokozera, C. Vesa, R.O. Wilson, M. Zheng, A. Kortyna, D.A. Tate, Radiative lifetime measurements of rubidium Rydberg states. J. Phys. B **43**, 015002 (2010)
27. K.J. Weatherill, A CO2 Laser Lattice Experiment for Cold Atoms. PhD thesis, Department of Physics, Durham University, 2007
28. E.L. Raab, M. Prentiss, A. Cable, S. Chu, D.E. Pritchard, Trapping of neutral sodium atoms with radiation pressure. Phys. Rev. Lett. **59**, 2631 (1987)
29. C.J. Foot, *Atomic Physics* (OUP, Oxford, 2005)
30. A.M. Steane, C.J. Foot, Laser cooling below the doppler limit in a magneto-optical trap. Europhys. Lett. **14**, 231 (1991)
31. D.A. Steck, Rubidium 85 D line data (2008), http://steck.us/alkalidata/rubidium85numbers.pdf
32. D.A. Steck, Rubidium 87 D line data (2007), http://steck.us/alkalidata/rubidium87numbers.pdf
33. D.J. McCarron, S.A. King, S.L. Cornish, Modulation transfer spectroscopy in atomic rubidium. Meas. Sci. Technol. **19**, 105601 (2008)
34. K.L. Corwin, Z.-T. Lu, C.F. Hand, R.J. Epstein, C.E. Wieman, Frequency-stabilized diode laser with the Zeeman shift in an atomic vapor. Appl. Opt. **37**, 3295 (1998)
35. U. Volz, H. Schmoranzer, Precision lifetime measurements on alkali atoms and on helium by beam-gas-laser spectroscopy. Phys. Scr. **T65**, 48 (1996)
36. C.G. Townsend, N.H. Edwards, C.J. Cooper, K.P. Zetie, C.J. Foot, A.M. Steane, P. Szriftgiser, H. Perrin, J. Dalibard, Phase-space density in the magneto-optical trap. Phys. Rev. A **52**, 1423 (1995)
37. A. Arnold, Preparation and Manipulation of an ^{87}Rb Bose-Einstein Condensate. PhD thesis, University of Sussex, 1999
38. P.D. Lett, W.D. Phillips, S.L. Rolston, C.E. Tanner, R.N. Watts, C.I. Westbrook, Optical molasses. J. Opt. Soc. Am. B **6**, 2084 (1989)
39. J. Dalibard, C. Cohen-Tannoudji, Laser cooling below the Doppler limit by polarisation gradients: simple theoretical models. J. Opt. Soc. Am. B **6**, 2023 (1989)
40. C.D. Wallace, T.P. Dinneen, K.Y.N. Tan, A. Kumarakrishnan, P. Gould, J. Javanainen, Measurements of temperature and spring constant in a magneto-optical trap. J. Opt. Soc. Am. B **11**, 703 (1994)
41. T.G. Walker, M. Saffman, Zeros of Rydberg–Rydberg förster interactions. J. Phys. B **38**, S309 (2005)

42. T.G. Walker, M. Saffman, Consequences of Zeeman degeneracy for the van der Waals blockade between Rydberg atoms. Phys. Rev. A **77**, 032723 (2008)

43. B.H. Bransden, C.J. Joachain, *Physics of Atoms and Molecules* (Longman Scientific and Technical, London, 1983

44. A.L. Migdall, J.V. Prodan, W.D. Phillips, T.H. Bergeman, H.J. Metcalf, First observation of magnetically trapped neutral atoms. Phys. Rev. Lett. **54**, 2596 (1985)

45. A. Reinhard, T. Cubel Liebisch, B. Knuffman, G. Raithel, Level shifts of rubidium Rydberg states due to binary interactions. Phys. Rev. A **75**, 032712 (2007)

46. R.P. Abel, A.K. Mohapatra, M.G. Bason, J.D. Pritchard, K.J. Weatherill, U. Raitzsch, C.S. Adams, Laser frequency stabilization to excited state transitions using electromagnetically induced transparency in a cascade system. Appl. Phys. Lett. **94**, 071107 (2009)

47. R.W.P. Drever, J.L. Hall, F.V. Kowalski, J. Hough, G.M. Ford, A.J. Munley, H. Ward, Laser phase and frequency stabilization using an optical resonator. Appl. Phys. B **31**, 97 (1983)

48. K. Singer, S. Jochim, M. Mudrich, A. Mosk, M. Weidemüller, Low-cost mechanical shutter for light beams. Rev. Sci. Inst. **73**, 4403 (2002)

49. I.G. Hughes, T.P.A. Hase, *Measurements and Their Uncertainties: A Practical Guide to Modern Error Analysis* (OUP, Oxford, 2010)

50. P. Siddons, C.S. Adams, C. Ge, I.G. Hughes, Absolute absorption on rubidium D lines: comparison between theory and experiment. J. Phys. B **41**, 155004 (2008)

51. J. Ye, S. Swartz, P. Jungner, J.L. Hall, Hyperfine structure and absolute frequency of the ^{87}Rb $5P_{3/2}$ state. Opt. Lett. **21**, 1280 (1996)

Chapter 7
Results

Rydberg EIT in thermal samples has already demonstrated coherent optical detection [1] and excitation [2] of the Rydberg states. In addition, the large polarisability of the Rydberg states have been exploited to control the properties of the probe field to create an optical switch [3] and a giant electro-optic effect that is 10^6 times larger than a typical nitrobenzene Kerr lens [4]. However, observation of dipole blockade in room temperature samples is challenging.

Using the apparatus described in the previous chapter, EIT experiments are performed on a cold atomic sample to look for evidence of the cooperative effects arising from the dipole–dipole interactions. Data are presented in the following sections for a range of principal quantum numbers, demonstrating two distinct regimes of behaviour. At low n ($\lesssim 26$), the interactions are weak and the superradiant broadening dominates over the level shifts. For the high n states (~ 60) the ensemble is blockaded, and the resulting non-linearity is characterised as a function of probe power and density for both attractive and repulsive interactions.

7.1 Low-n EIT

For the initial experiments, spectroscopy is performed on the low-n states ($n \lesssim 26$) to demonstrate EIT as a non-destructive probe of the Rydberg energy states. The probe beam is collimated and passed through an aperture to give a beam of approximately uniform intensity with a radius of 0.75 mm, ensuring the entire atom cloud is illuminated. The coupling laser was collimated to a $1/e^2$ waist of 0.8 mm with a peak power of 100 mW. ^{85}Rb atoms are loaded in the MOT for 5 s to maximise the cloud size, and hence optical depth along the probe beam, giving 10^8 atoms with a peak density of 10^{10} cm^{-3} and a cloud with a radius of $\sigma_r = 0.7$ mm. Following the molasses, atoms are prepared in the ($F = 3, m_F = 3$) ground state using bright state optical pumping with the probe laser for a pulse length of 10–100 µs dependent upon

J. D. Pritchard, *Cooperative Optical Non-Linearity in a Blockaded Rydberg Ensemble*, Springer Theses, DOI: 10.1007/978-3-642-29712-0_7,
© Springer-Verlag Berlin Heidelberg 2012

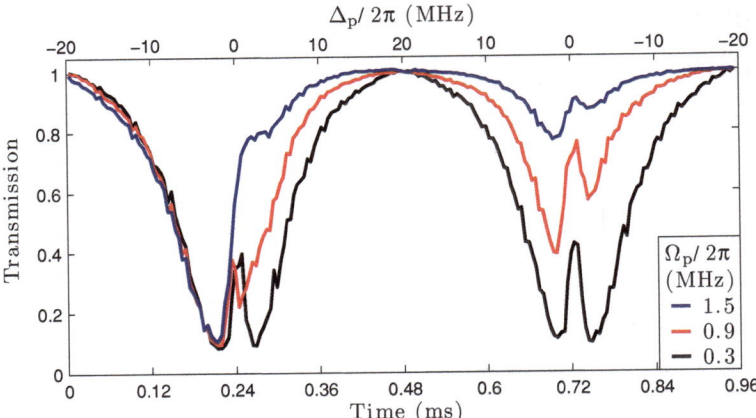

Fig. 7.1 EIT data for $22D_{5/2}$ for probe powers of $200\,\text{nW}$ to $3.6\,\mu\text{W}$ corresponding to Rabi frequencies of $\Omega_p/2\pi = 0.3{-}1.5\,\text{MHz}$. At low power a narrow EIT spectra is observed, however as the probe power is increased a non-linear loss mechanism is observed

the probe power. The atoms are then probed using an NPN probe laser frequency sweep in a period of $960\,\mu\text{s}$.

7.1.1 Weak Probe Spectroscopy

Figure 7.1 shows exemplary transmission spectra for the $22D_{5/2}$ Rydberg state recorded using $75\,\text{mW}$ of coupling laser power at a range of probe powers. Considering first the weak-probe spectrum for $\Omega_p/2\pi = 0.3\,\text{MHz}$, this shows two symmetric scans with a narrow transparency window appearing in the centre of the $5P_{3/2}$ $F' = 4$ absorption feature. These data represent a single run of the experiment from which it is possible to determine both the dephasing rate γ_{gr} of the two-photon transition and the effective coupling Rabi frequency Ω_c, with values of 0.30 ± 0.05 and $3.6 \pm 0.2(\times 2\pi)\,\text{MHz}$ respectively. The lifetime of the $22D_{5/2}$ state is $8\,\mu\text{s}$ [5] corresponding to a natural linewidth of $\Gamma_r/2\pi = 20\,\text{kHz}$, an order of magnitude less than the measured dephasing rate. This dephasing, $\gamma_{gr} = \Gamma_r/2 + \gamma_{\text{rel}}$, is therefore dominated by the relative two-photon laser linewidth of $300\,\text{kHz}$, which limits the frequency resolution of the EIT to a FWHM of $\gamma_{\text{EIT}}/2\pi = 1\,\text{MHz}$ for these parameters. Subsequent improvements of the laser stabilisation have reduced this relative linewidth to give $\gamma_{\text{rel}}/2\pi \sim 100\,\text{kHz}$, allowing resonances with FWHM of $600\,\text{kHz}$ to be observed for $26D_{5/2}$ [6].

EIT therefore provides a non-destructive probe of the Rydberg state energies without actually transferring population into the Rydberg state. The sub-MHz resolution is comparable to spectroscopy performed on an isolated single atom [7] and better than has been obtained previously in other experiments using Rydberg ensembles

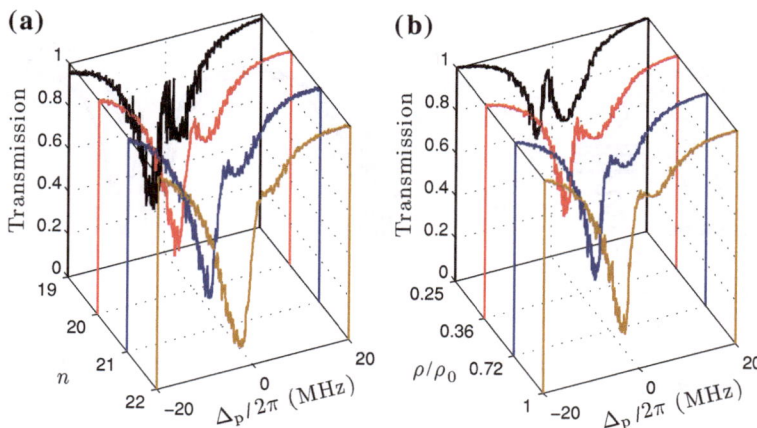

Fig. 7.2 a EIT spectra for $n = 19 - 22D_{5/2}$ with uniform coupling Rabi frequency of $\Omega_p/2\pi = 1.5\,\text{MHz}$ **b** transmission as a function of density for $26D_{5/2}$

[8–11] which suffer from interaction-induced broadening. Rydberg EIT is thus suitable for applications in electrometry [3], and has been used to measure electric fields close to surfaces with a sensitivity of 0.1 V/cm [12].

7.1.2 Strong Probe Regime

Outside of the weak-probe limit, the Rydberg component of the dark-state $|D\rangle$ increases proportional to $\sim \Omega_p/\Omega_c$. Neglecting the effects of interactions, this should result in a reduced absorption in the wings of the EIT resonance due to population transfer out of the ground state. Returning to the data presented in Fig. 7.1, at high probe power a very asymmetric transmission profile is observed, which has a pronounced enhancement in transmission at the start of the two-photon resonance. This feature is associated with significant loss of atoms, as can be seen by the reduction of optical depth in the reverse scan across resonance. For this second scan where the density is lower, there is a recovery of the EIT even for $\Omega_p/2\pi = 1.5\,\text{MHz}$ which has no distinguishable resonant feature in the first scan.

To explore this effect further, EIT spectra are taken at $\Omega_p/2\pi = 1.5\,\text{MHz}$ for a range of principal quantum numbers and for different densities. To maintain a constant value of Ω_c across the datasets, the power in the coupling laser is scaled proportional to $n^{*-3/2}$ for each state, matching the scaling of the transition dipole moment from $5P_{3/2}$ to $n\ell$ (see Sect. 2.3.3).

Transmission data for the first scan across the EIT resonance are presented in Fig. 7.2 as a function of (a) quantum number and (b) density. In (a), the spectra evolve smoothly from a well resolved EIT resonance with slight asymmetry at $n = 19D_{5/2}$ to the pronounced loss at $n = 22D_{5/2}$, as seen in Fig. 7.1. This loss is

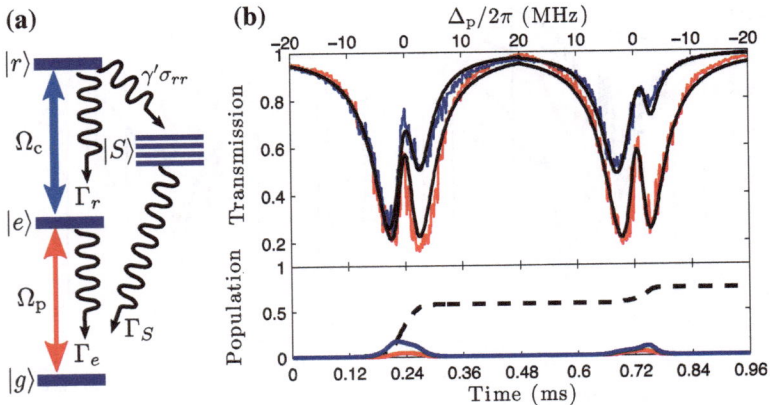

Fig. 7.3 Density-dependent loss model. **a** Level scheme for model, with decay into state $|S\rangle$ proportional to the Rydberg state density σ_{rr}. **b** EIT data for $19D_{5/2}$ for $\Omega_p/2\pi = 0.5$ (*red*) and 1.5 MHz (*blue*). *Black lines* show results obtained by fitting with four-level model. Below, the populations of the Rydberg state $\sigma_{rr} \times 100$ (*solid*) and the population of the reservoir state $|S\rangle$ for the strong probe (*dashed*) are plotted, showing $> 75\%$ population transfer into $|S\rangle$

observed consistently for states up to $26D_{5/2}$, which can be seen from the highest density trace in (b). Due to the wavelength range of the coupling laser it was not possible to excite atoms below $n = 19$ to see if the symmetry of the EIT recovers completely. In (b), the atom cloud is heated by optically pumping for increasing durations on the closed probe transition before performing the EIT on $26D_{5/2}$. This reduces the ground state density in the system and increases the sample size. Spectra are taken for fixed probe and coupling laser parameters, revealing almost complete recovery of the EIT lineshape as the density is reduced. Comparison of (a) and (b) reveal an almost indistinguishable evolution in the spectra, which implies a density dependent interaction that causes rapid depopulation of the Rydberg states. The timescale associated with this loss is much faster than the expected lifetime of the Rydberg states, which are of the order of 10 μs at 300 K [5].

The non-linear density scaling can be modelled phenomenologically using a modified set of optical Bloch equations from Sect. 4.1, in which an additional level $|S\rangle$ is introduced, shown in Fig. 7.3a. This state $|S\rangle$ acts as a reservoir state which the Rydberg atoms can decay into, removing them from the three-level EIT system to consistently reproduce loss. To enable spontaneous emission from $|r\rangle$ into both $|S\rangle$ and $|e\rangle$, a branching ratio ϵ is introduced to control the weighting of these two decay channels. The density dependence is included in the Lindblad operator $\mathcal{L}(\sigma)$ (Eq. 4.9) by setting the operator for the decay channel from $|r\rangle$ into $|S\rangle$ equal to $C_{s'} = \sqrt{(1-\epsilon)\Gamma_r + \gamma'\sigma_{rr}}|S\rangle\langle r|$, where the first term accounts for spontaneous emission from $|r\rangle$ and the second term is an interaction-induced loss rate proportional to the Rydberg state population, σ_{rr}. Correspondingly, the decay from $|r\rangle$ to $|e\rangle$ becomes $C_r = \sqrt{\epsilon\Gamma_r}|e\rangle\langle r|$. Finally, the decay from $|S\rangle$ back to $|g\rangle$ is included using $C_s = \sqrt{\Gamma_s}|g\rangle\langle S|$. To calculate the transmission, the optical Bloch equations are solved in time for the full frequency sweep. Values for the coupling Rabi frequency, relative linewidths and laser detunings are determined from fitting the weak probe spectra, and the parameters ϵ, γ' and Γ_S are found by comparing the OBE model to the strong probe data.

Figure 7.3b shows EIT spectra for $19D_{5/2}$ at $\Omega_p/2\pi = 0.5$ and 1.5 MHz compared to the best-fit transmission predicted by the OBE model, where the only parameter changed in the calculation is the probe Rabi frequency. The model is relatively insensitive to the value of ϵ, whilst the decay back into $|g\rangle$ must occur very slowly, corresponding to an effective state lifetime $\tau_S > 1$ ms. The only significant parameter is therefore γ', adjusted to give a peak loss rate of $960 \times 2\pi$ kHz for the strong probe data. The model shows very good agreement with the experimental traces, reproducing both the EIT spectra and the loss observed in the second scan. In the lower panel, the Rydberg atom populations are plotted showing a peak population around 2 % due to the rapid loss with increasing population, and showing greater than 70 % of the initial population is transferred into the reservoir state $|S\rangle$. Attempts to fit the higher n-states show good qualitative agreement with the lineshape of the first scan requiring an increased value of γ', however it is not possible to consistently reproduce the sharp loss feature and predict the correct density in the second scan.

The model verifies the spectra are caused by a density-dependent loss from the Rydberg state. Repeating the experiments for nearby $S_{1/2}$ states yields very similar results to those of Fig. 7.2, showing the effect is independent of the attractive or repulsive dipole–dipole interactions. This is expected, as the interaction shift at the average interatomic separation of 2 μm for the $26D_{5/2}$ state is only around 250 kHz, too small to observe blockade effects. This weak dipole interaction also rules out mechanisms such as dipole–dipole energy transfer [13]. The most likely explanation is therefore superradiant cascade from the Rydberg state. As discussed in Sect. 5.2, the condition for observing superradiant decay is $kR \simeq 1$, where k is the wavevector of the decay channel and R is the sample size. For the $nD_{5/2}$ states, the longest wavelength decay channel is via the $(n-2)F$ states, which approximately twice as big as for decay via the nP states. The decay wavelengths are $\lambda_{DF} = 0.35$ and 1.0 mm for 19 and $26D_{5/2}$ respectively, which is comparable to the MOT diameter of 1.4 mm for $n = 26$. The result is a geometric enhancement in the superradiant decay, with the cooperativity parameter of Eq. 5.9 changing by two orders of magnitude from $C = 10^{-5}$ to 10^{-3} from 19 to 26, which can be seen in the evolution of the spectra for increasing n. Further evidence for cooperativity is provided by the data in Fig. 7.2. In (a) kR is varied by changing wavelength to give a ratio of $0.57/0.35 = 1.6$ between $19D_{5/2}$ and $22D_{5/2}$, whilst in (b) kR is varied by changing the density by a factor of 4, and hence R by a ratio of $\sqrt[3]{4} = 1.6$. This scaling explains the similarity between the two datasets, and shows this superradiant decay mechanism gives the EIT a reproducible and characteristic lineshape. These findings are consistent with more detailed studies of superradiance in cold Rydberg gases for $n = 20 - 30$ [14, 15], however without ion detection it is not possible to conclusively verify the population is cascading down to lower n states, as the mm-wave emission cannot be detected.

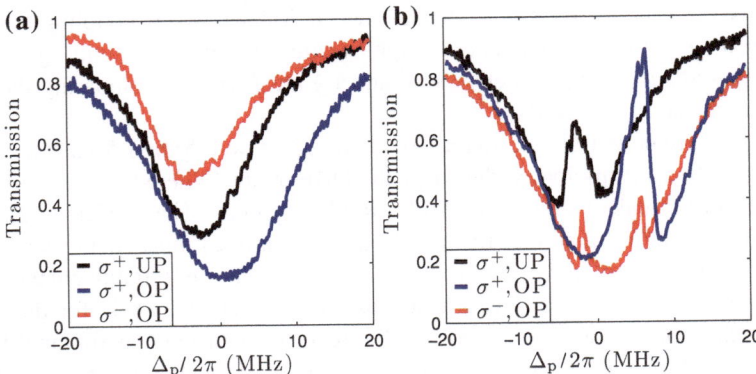

Fig. 7.4 Importance of optical pumping in EIT spectra. **a** Probe transmission for $\sigma^{\pm}$ transitions with no optical pumping (UP) or dark-state pumping into $(F = 2, m_F = 2)$ (OP). **b** $44D_{5/2}$ EIT spectra for different coupling laser polarisations with a σ^{+} probe beam

7.1.3 Summary

As discussed previously, the superradiant decay rate scales as n^{*-5} compared to n^{*11} for the dipole–dipole interactions. To observe cooperative behaviour due to the energy shift rather than enhanced broadening, it is therefore necessary to use states with higher n. These initial results, however, show that in the weak probe regime EIT provides a non-destructive, state-selective probe of the Rydberg states which may be useful for high resolution spectroscopy.

7.2 High n EIT : Optical Pumping and Polarisation

Extending the experiments to higher n states requires tighter focusing of the coupling laser to give a sufficiently high Rabi frequency to observe EIT. The probe laser also needs to have a smaller focus than the coupling laser to ensure the condition $\Omega_c > \Omega_p$ is achieved across the whole cloud. Using the adjustable focal length fibre collimators, the probe and coupling lasers are therefore focused to $1/e^2$ radii of 160 and $215\,\mu$m respectively. Another important change in the experiment procedure is to use ^{87}Rb to enable the dark-state optical pumping scheme to be used, as detailed in Sect. 6.2. To illustrate the importance of efficient optical pumping, Fig. 7.4a shows absorption data recorded for a 1 s MOT load using a 10 nW probe ($\Omega_p/2\pi = 0.5$ MHz) polarised to drive either σ^{+} or σ^{-} transitions. The middle (black) trace shows transmission for a σ^{+} probe beam without any optical pumping stage or bias field, giving a peak absorption of 70 %. If the optical pumping step is used however, the absorption now increases to 85 %, with the shift in frequency due to the 2 G bias field causing a Zeeman shift of the $(F = 2, m_F = 2)$ state. If the probe beam polarisation is now

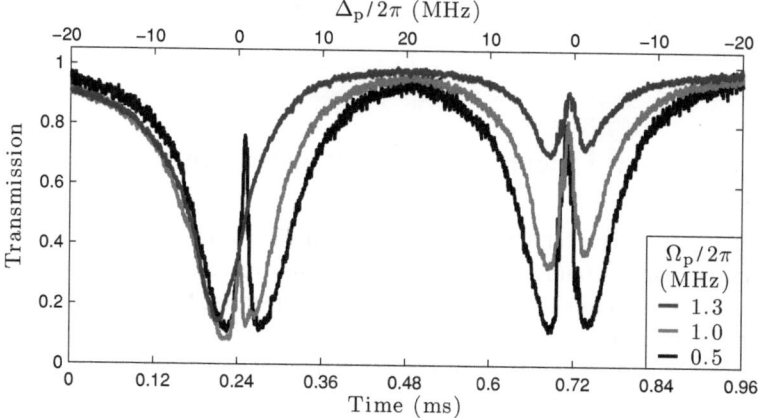

Fig. 7.5 EIT spectra for $46D_{5/2}$ showing the characteristic signatures of super-radiant loss observed at low n

reversed to σ^-, the absorption is reduced and Zeeman shifted to negative frequency as atoms are pumped across to the ($F = 2$, $m_F = -2$) state during the scan sequence, giving a kink in the transmission curve at $\Delta_p/2\pi = -10$ MHz. The optical pumping therefore significantly enhances the optical depth in the cold atom sample, with all atoms on the closed optical transition. The MOT load time is 1 s in all subsequent data.

Optical pumping is also important for enhancing the visibility of the EIT resonance. Figure 7.4b shows EIT on the $44D_{5/2}$ state with the probe beam polarised to drive a σ^+ transition, using the same beam power as in (a). If the atoms are not optically pumped, the σ^+ coupling laser gives a broad transmission window on resonance with around 20 % change in transmission. If the atoms are optically pumped however, the EIT is dramatically enhanced with 90 % transmission on resonance. This is because the σ^+–σ^+ configuration drives two closed transitions up to the $D_{5/2}$ state. If the coupling laser polarisation is reversed to drive the σ^- transition, the EIT is suppressed as the laser is now driving the weakest optical transition. This trace shows an additional resonance at the same frequency as the un-pumped data due to the Zeeman shift of the Rydberg state combined with imperfect optical pumping, which allows the atoms remaining in $m_F < 2$ to have a stronger coupling to the Rydberg state. Repeating this for the $S_{1/2}$ states, the EIT can be turned off completely when the coupling laser drives a σ^+ transition as this violates the selection rule for excitation from $5P_{3/2}$.

Having optimised the optical pumping to maximise the optical density and coupling Rabi frequency, spectra are taken for increasing probe powers to look for evidence of an optical non-linearity. Results are presented in Fig. 7.5 for the $46D_{5/2}$ state, which shows narrow EIT spectra at low power but a sharp loss feature at high power, very similar to the lineshape observed in Fig. 7.1. This shows that even at $n = 46$ the superradiant loss mechanism is still dominating over dipole–dipole interactions.

7.3 Optical Non-Linearity due to Attractive Interactions

Continuing to focus on the $nD_{5/2}$ states due to their stronger coupling to the $5P_{3/2}$ state relative to $nS_{1/2}$ (and hence larger transparency on resonance), states of higher principal quantum number were studied [16]. As discussed in Sect. 3.5, the $nD_{5/2}$ states experience attractive long-range interactions which causes atoms to be accelerated together and collide to form ions on a timescale of around $10\,\mu s$ [17]. It is therefore necessary to consider temporal effects due to atomic motion.

7.3.1 Temporal Dependence

The motional dynamics only play a role in EIT in the strong probe regime, when the resonant dark state contains a non-zero Rydberg fraction. Using the optical Bloch equations for a single, non-interacting atom it is possible to calculate the probe susceptibility and Rydberg population σ_{rr} during the probe frequency ramp as a function of scan speed. Figure 7.6a shows the results calculated using $\Omega_p/2\pi = 0.9\,\mathrm{MHz}$, $\Omega_c/2\pi = 2.4\,\mathrm{MHz}$ and $\gamma_{rel}/2\pi=200\,\mathrm{kHz}$ to match experiment parameters. This illustrates the Rydberg population has an excitation bandwidth comparable to the width of the EIT transmission window, which for these parameters corresponds to a FWHM of $\gamma_{EIT}/2\pi = 1.2\,\mathrm{MHz}$. The Rydberg states are therefore populated even in the wings of the EIT resonance. These results represent the steady-state solution for the system, being independent of the frequency scan parameters for scan speeds up to 1 GHz/ms. For higher speeds, the spectrum is distorted by transient effects as the probe frequency is changing on a timescale comparable to the time required to establish the dark state coherence. Thus, in the absence of any motion or interactions the spectra should be independent of probe scan speeds below this rate.

To test this steady-state assumption for the interacting system, transmission is recorded for the $58D_{5/2}$ state at a density of $\rho = 0.9\pm0.1 \times 10^{10}\,\mathrm{cm}^{-3}$ for a positive scan across the EIT resonance at a range of scan speeds, shown in Fig. 7.6b. For the slowest speed of 50 MHz/ms, the laser scans across the EIT resonance in $60\,\mu s$, however the EIT feature is poorly resolved. Instead, the resonance appears broadened, starting at $\Delta_p/2\pi = -2\,\mathrm{MHz}$ on the edge of the two-photon transition and showing a rapid change in transmission at $\Delta_p/2\pi \sim 0.5\,\mathrm{MHz}$ as atoms are lost from the sample, seen from the narrowing of the width of the probe absorption feature. This lineshape is consistent with initial loss due to ionisation of the close-spaced anti-blockade states which are red-shifted due to the attractive interactions, and ionise rapidly. These residual ions lead to a Stark-shift across the cloud, broadening and suppressing the EIT. The large loss at $\Delta_p/2\pi \sim 0.5\,\mathrm{MHz}$ occurs approximately

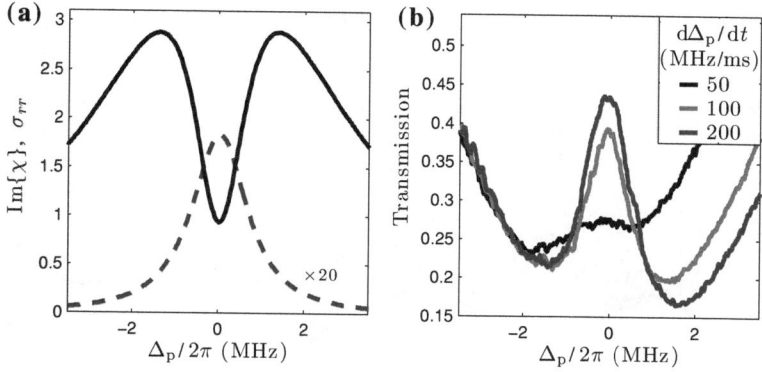

Fig. 7.6 Temporal effects in EIT. **a** Probe susceptibility (*solid line*) and Rydberg population (*dashed line*) calculated using the single-atom OBEs from Sect. 4.1 show the Rydberg state population has the same frequency width as the EIT resonance. The results are independent of scan duration for scan speeds up to up to 1 GHz/ms. **b** Spectra recorded for $58D_{5/2}$ at different scan speeds for $\Omega_p/2\pi = 0.9$ MHz. For the slowest scan (50 MHz/ms) there is broadening and loss consistent with ionisation, however the faster scans show no evidence of ionisation and are consistent

10 µs after the exact two-photon resonance, which could be due to the ionisation of the long-range pair states. For higher scan speeds however, the spectra become symmetric with approximately constant transmission on the EIT resonance, showing the ion-induced loss is suppressed. The data below are all taken using a scan speed of 80 MHz/ms, fast enough that there is no evidence of loss or asymmetry across the EIT resonance.

7.3.2 EIT Suppression

Having set the probe scan speed to avoid any obvious asymmetry or loss in the scan across resonance, the experiment is repeated at an increased density of $\rho = 1.6 \pm 0.2 \times 10^{10}$ cm^{-3} to look for evidence of cooperativity due to dipole–dipole interactions. The resulting spectra are shown in Fig. 7.7a, which shows a strong, symmetric suppression of the resonant transmission for high probe power as expected from the model in Sect. 5.4. Whilst this appears to show a cooperative suppression, it is necessary to consider the effect of a small ion fraction within the atom cloud.

Ionisation is an incoherent mechanism that leads to a random distribution of charges in the system. As the Rydberg states have very large polarisabilities, these random electric fields can dominate over the quantisation axis provided by the bias field, projecting the atom into a random $|m_j|$ state. Using the fit parameters from Table 2.2, the scalar polarisabilities of the $58D_{5/2}$ states are $\alpha_0 = -137$, 111 and 607 MHz/(V/cm)^2 for $|m_j| = 1/2$, 3/2 and 5/2 respectively. These states therefore experience shifts of different sign and strength, which means the signature of ion-

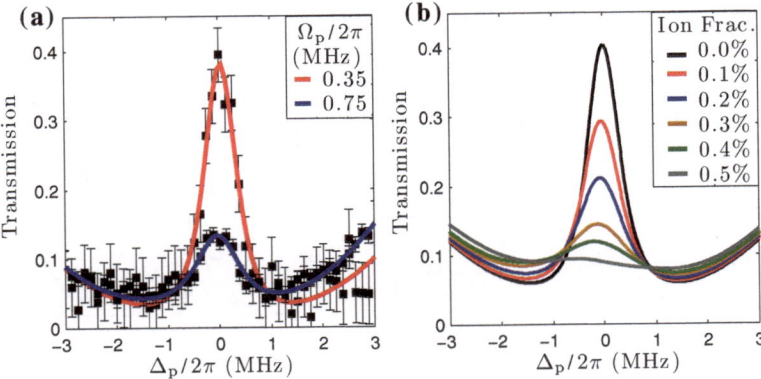

Fig. 7.7 $58D_{5/2}$ EIT Suppression. **a** Transmission data showing strong suppression of the EIT for a strong probe beam as expected from cooperative effects. **b** Simulated lineshape to model the effects of ionisation. Due to the difference in sign of α_0 for the different $|m_j|$ states, the EIT is not shifted but instead suppressed by ions, with the 0.3 % ion fraction comparable to suppression data in **a**

isation is not simply a net detuning of the two-photon resonance. To give an idea of the effect of this mixture of shifts on the lineshape, a Monte-Carlo model was used to randomly pick 10^5 atoms from a uniform density distribution and choose a fixed fraction of them as ions. For each of the remaining atoms, the total electric field due to all surrounding ions is calculated and the Stark-shift calculated using a randomly assigned $|m_j|$ state. The susceptibility is then found using the weak-probe formula of Eq. 4.20 and the total transmission profile obtained by summing over the susceptibility of each atom. The results are presented in Fig. 7.7b, calculated using $\Omega_c/2\pi = 2.6\,\mathrm{MHz}$ and $\gamma_{\mathrm{rel}}/2\pi = 200\,\mathrm{kHz}$ to match the values obtained from fitting the weak-probe spectrum in (a). This shows that for only a 0.3 % ion-fraction the EIT is suppressed to a similar level to that observed in the experiment, making the effect of ionisation and blockade difficult to distinguish.

One caveat to using this model for comparison to data is that it makes several assumptions; firstly, that the ions are present for the full duration of the scan rather than being created dynamically, and secondly, that there is an even distribution of the atoms into each of the $|m_j|$ components. In addition, the blockade mechanism should prevent excitation of closely-spaced pair states, suppressing the ion creation during the scan across the two-photon resonance. For the $58D_{5/2}$ $m_{j1} = m_{j2} = 5/2$ pair state the interaction strength is calculated as $C_6 = 150\,\mathrm{GHz\,\mu m^6}$, which for the EIT linewidth of $\gamma_{\mathrm{EIT}}/2\pi = 0.8\,\mathrm{MHz}$ corresponds to a blockade radius of $R_b \sim 8\,\mu\mathrm{m}$. For atoms with this initial separation, the timescale for collisions is tens of microseconds, longer than the time taken for the probe laser to scan across the EIT bandwidth.

As the ion yield cannot be directly measured for these experiments, it is not possible to completely discount the effect of ionisation in the observed suppression data. However, the EIT remains symmetric with no shift or broadening of the two-photon resonance over a wide range of probe powers, suggesting the dipole–dipole interactions are the dominant mechanism. It is still interesting though to consider the

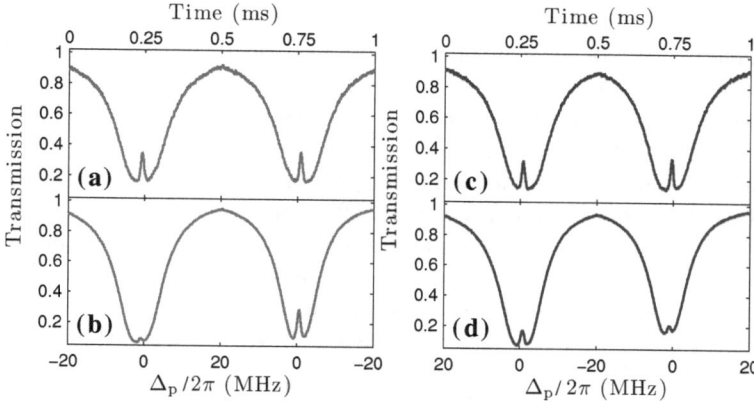

Fig. 7.8 Transmission spectra for probe Rabi frequencies of $\Omega_p/2\pi = 0.3$ MHz **a, c** and 0.9 MHz **b, d**. Data are recorded for NPN **a, b** and PNP **c, d** frequency sweeps. The weak probe data of **a** and **c** shows symmetric EIT spectra with no evidence of hysteresis. For the strong probe data however, the NPN scan **b** has enhanced suppression relative to the PNP spectrum in **d**, with more loss on the second scan

non-linear response of the system due to a process such as ionisation. Whilst this is an incoherent effect, the long-range Coulomb field could be used to Stark-shift the EIT off-resonance to create an ion-blockade [18] which could switch the medium from transparent to opaque, making a sensitive ion detector.

7.3.3 Frequency Dependence of the EIT Suppression

An additional handle that can be used to assess the importance of dynamical effects in the EIT is to perform the spectroscopy with a reversed frequency scan direction. The reason this should make a difference is due to the presence of the red-detuned anti-blockaded pair states, which correspond to resonant excitation of atoms with separation smaller than R_b. If these states are important in the evolution of the EIT at high powers, then the spectra will show signs of hysteresis dependent upon whether the probe laser scans across them before or after reaching the two-photon resonance.

Figure 7.8 shows transmission data recorded at probe Rabi frequencies of $\Omega_p/2\pi = 0.3$ MHz in (a) and (c), and 0.9 MHz in (b) and (d) for NPN and PNP frequency sweeps across the EIT resonance at a density of $1.6 \pm 0.2 \times 10^{10}$ cm^{-3}. For the weak probe data, symmetric EIT spectra are observed for both frequency sequences, with no evidence of hysteresis between the first and second scan across resonance. This symmetry is due to the Rydberg state not being populated for low probe Rabi frequencies. For the strong probe NPN transmission spectrum in (b), there is again a significant suppression of the EIT resonance in the first scan across resonance as seen above in Fig. 7.7a.

However, in the reverse scan the EIT appears to recover almost completely, with a slight reduction in the optical depth suggesting some atoms have been lost during the scan sequence. Comparing this with the strong probe PNP data in (d), this also shows a suppression of the EIT on the first scan compared to the weak probe regime, though much less than is seen in (b). On the second scan though, the suppression is significant, with a larger reduction in the ground state density after the first scan. These figures clearly illustrate a hysteresis effect in the EIT spectroscopy, with the suppression maximised when scanning from negative to positive detunings across the two-photon resonance. An additional parameter that can be found from the data is the detuning of the two-photon resonance. For each set of parameters, the two-photon resonance is compared across 20 data sets. No systematic shift is observed above the $\pm 300\,\text{kHz}$ variation measured in the weak probe limit, which arises due to the fluctuations in the frequency of the coupling laser.

These results show the EIT suppression is sensitive to the scan direction, with the behaviour consistent with pair excitation. For the PNP sequence in (d), the laser scans across the two-photon resonance, with suppression in the transmission appearing due to blockade of the resonant dark state. The laser then becomes negatively detuned, and can resonantly excite the anti-blockaded pair states which ionise, leading to the observed loss in the second scan. For the NPN data however, the laser is initially red-detuned and can therefore excite the closely-spaced pair states. These pair states will ionise rapidly, creating a residual ion fraction in the cloud on the two-photon resonance which would enhance the suppression of the EIT, as seen in Fig. 7.7b. This explains the difference in the first scan for the PNP scan compared to NPN. Following ionisation, these short-range pairs will be lost from the probe region, modifying the nearest-neighbour distribution in the cloud. The result is less atoms in each blockade sphere, which leads to a recovery of the EIT in the second scan across resonance.

Whilst this analysis is speculative, due to the lack of a time-resolved ion signal to accompany the transmission spectra, the results are similar to direct studies of Rydberg population in which the ion yield due to the anti-blockade states is seen to be enhanced for excitation of the attractive $D_{5/2}$ states with a red-detuned laser relative to blue-detuned excitation [17, 19]. Despite the exact mechanism for the strong suppression seen for the NPN sequence being unclear, it is still useful to characterise the optical non-linearity resulting from these interactions.

7.3.4 Optical Non-Linearity

To make a quantitative measurement of the optical non-linearity arising from the suppression of the EIT resonance, transmission data are recorded for a range of probe powers and densities, using a variable depump time to systematically change the ground-state density without changing the cloud size, as described in Sect. 6.2.1. Spectra are recorded for both NPN and PNP scan sequences, from which the transmission on the two-photon resonance is found. The transmission, T, is converted to the imaginary part of the susceptibility using Eq. 4.23a to give $\chi_I = -\log_e(T)/k\ell$,

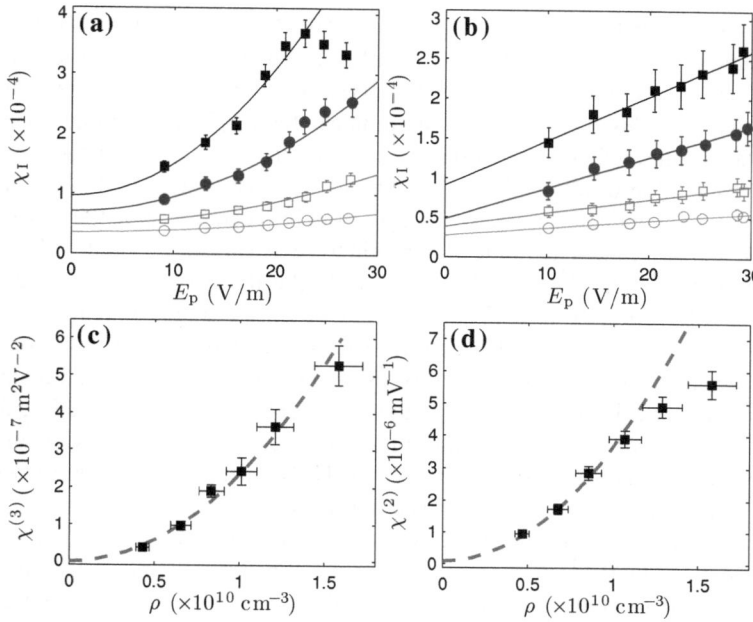

Fig. 7.9 Resonant susceptibility as a function of probe electric field, E_p, and density, ρ. **a** and **b** show data for NPN and PNP scan directions respectively, taken for $\rho = 0.4$ (○) 0.7 (□) 1.0 (●) 1.6 (■) $\times 10^{10}\,\mathrm{cm}^{-3}$. Data in **a** and **b** is fitted to third- and second-order non-linearities respectively, and the resulting density dependence plotted in **c** and **d**. Both $\chi^{(2)}$ and $\chi^{(3)}$ display a quadratic density scaling, consistent with pair-wise interactions

where k is the probe wavevector and ℓ is the optical path length through the cloud, measured to be $\ell = 0.9 \pm 0.1\,\mathrm{mm}$. For each set of parameters, the susceptibility is calculated for 20 repeats, and then a weighted average taken to give the final value. This is plotted against the probe electric field, E_p, to look for a non-linear scaling.

Results are shown in Fig. 7.9a,b measured for the NPN and PNP scan sequences respectively for a range of densities. Immediately obvious is the difference in the non-linear scaling for the different scan directions. For NPN, there is clear evidence of a third-order non-linearity in agreement with the theoretical prediction of [20]. This non-linearity saturates for probe powers above 20 V/m at the peak density of $1.6 \times 10^{10}\,\mathrm{cm}^{-3}$. Increasing the probe power further, there is now an increase in the transmission rather than suppression, most likely caused by loss as the blockade mechanism breaks down and ionisation becomes dominant. The data are fit to the function $\chi_I = \chi^{(1)} + \chi^{(3)} E_p^2$, from which the third-order susceptibility can be measured. The peak value of $\chi^{(3)} = 5.3 \pm 0.4 \times 10^{-7}\,\mathrm{m^2\,V^{-2}}$ represents a very large non-linear scaling for an atomic ensemble, comparable in magnitude to the slow-light experiments in a BEC [21] performed at densities two orders of magnitude higher than are used here. However, the PNP data in (b) are not consistent with a third-order scaling, but instead agree well with a second-order scaling. The data are therefore fit using $\chi_I = \chi^{(1)} + \chi^{(2)} E_p$ to extract the value of $\chi^{(2)}$, which has a peak value of $\chi^{(2)} = 5.6 \pm 0.4 \times 10^{-6}\,\mathrm{mV}^{-1}$. Typically, this second order effect can only be seen in non-centrosymmetric crystals in which the oscillating electrons that form the dipoles experience an anharmonic potential [22]. The broken symmetry in crystals is an artefact of the collective interaction of the atoms in each unit cell, so the observation of a $\chi^{(2)}$ dependence here is still consistent with a cooperative effect where the optical response of a single atom is dependent upon the surrounding atoms. Comparing the magnitude obtained in the experiment to that of bulk crystals, which typically have $\chi^{(2)} \sim 10^{-10}\,\mathrm{mV}^{-1}$ [22], the Rydberg blockade mechanism gives a non-linearity 10^4 times larger than can be achieved in a crystalline medium.

The important signature of a cooperative optical non-linearity is a non-linear density dependence in the susceptibility, as increasing the number of atoms in each blockade sphere enhances the suppression of the EIT resonance. Figure 7.9c,d show the third- and second-order susceptibilities obtained from fitting the NPN and PNP data respectively as a function of density. The NPN third-order susceptibility shows very good agreement with a quadratic scaling across the full range of densities, whilst the second-order susceptibility initially agrees with a quadratic scaling but saturates above a density of $10^{10}\,\mathrm{cm}^{-3}$. These results are consistent with a cooperative non-linear mechanism, and agree with theoretical predictions for a quadratic density scaling obtained using a Monte-Carlo method to calculate the optical response of very large atom numbers [20]. The Monte-Carlo model also predicts a saturation in the quadratic density scaling around $10^{10}\,\mathrm{cm}^{-3}$, as seen in (d).

7.3.5 Summary

These observations of suppression for the attractive Rydberg states show clear evidence of cooperativity, reproducing the expected suppression of the EIT resonance and displaying a quadratic density dependence consistent with the blockade mechanism. The data also reveal some interesting dynamical properties which can be seen through the hysteresis between both the first and second scan across resonance. This is demonstrated by the recovery of the EIT in the second scan in Fig. 7.8b, and through the dependence upon scan direction. These effects are most likely related to ionisation of the short-range anti-blockaded states which are excited when the probe is negatively detuned. Without being able to detect the ion fraction explicitly, this data cannot be used as conclusive proof of the blockade-induced cooperative non-linearity described in Sect. 5.4. Rydberg EIT has, however, been demonstrated to give a very large third-order non-linearity, and it would be interesting for the exact mechanism to be verified in future studies. If ions are the cause of the suppression, it could create a very sensitive optical method for performing single ion detection.

The $D_{5/2}$ states have two additional properties that have not been exploited in this present work. Firstly, the anisotropic interactions mean that the interaction depends on the alignment of the dipoles [23], which could be used to control or tune the non-linearity. Secondly, the D pair states have small energy defects, as seen in Fig. 3.3, allowing tuning to $1/R^3$ resonant dipole–dipole interactions using a Förster resonance. This would enhance the non-linearity by increasing the size of the blockade radius, and hence the number of atoms contributing to the suppression.

An additional mechanism for suppression that has not been considered so far is the van der Waals dephasing discussed in Sect. 5.2, which leads to an in-homogeneous broadening of the EIT resonance due to the distribution of interaction strengths in the cloud. As the visibility of the EIT resonance is limited in the present setup, it is not possible to distinguish this broadening effect from suppression due to interactions. A number of changes are therefore needed to allow conclusive verification of the cooperative suppression due to interactions; namely a clear spectral signature of ionisation, and increased transparency on the two photon resonance.

7.4 Cooperativity due to Repulsive Interactions

7.4.1 Experiment Modifications

The main limitation in the analysis of the previous section arises from the different signs in the Stark shifts of the $|m_j|$ components of the Rydberg state, leading to a suppressed transmission without a clear shift in the resonance. This can be overcome by using the $S_{1/2}$ states [24], which have two important advantages over the D states. Firstly, there is only a single $|m_j|$ component for which the scalar polarisability $\alpha_0 > 0$, ensuring that all atoms experience a Stark-shift to negative detuning if there are ions present in the sample. Secondly, the atoms experience isotropic, repulsive dipole–dipole interactions, which significantly reduces the ionisation rate relative to the D states, as discussed in Sect. 3.5.

The other issue to address is the magnitude of the transparency on the two-photon resonance, which was limited by the weak coupling Rabi frequency for the $58D_{5/2}$ state. The coupling laser is therefore focused down to a $1/e^2$ radius of $66 \pm 3\,\mu$m, the smallest waist possible using the fibre collimator. As the Rabi frequency is inversely proportional to the waist, this gives a factor of ~ 3 enhancement in Ω_c relative to the previous experiments. In addition to maximising the coupling laser Rabi frequency, the probe beam was then focused to a $1/e^2$ radius of $12 \pm 0.2\,\mu$m. This ensures an approximately uniform coupling Rabi frequency across the probe beam to give the largest possible transparency for all atoms in the probe region. To achieve this tight waist, the setup was modified as shown in Fig. 7.10a, with the probe fibre output collimated to a $1/e^2$ waist of 3.4 mm which was then focused using a 15 cm focal length doublet outside the chamber. Since the output probe beam is highly divergent, it is necessary to re-collimate it with another lens after the chamber. This required moving the optical pumping fibre to co-propagate with the probe beam, making a 1:1 telescope that uses the doublet as the second lens to ensure the optical pumping beam is not focused inside the chamber. After the chamber, a second telescope is

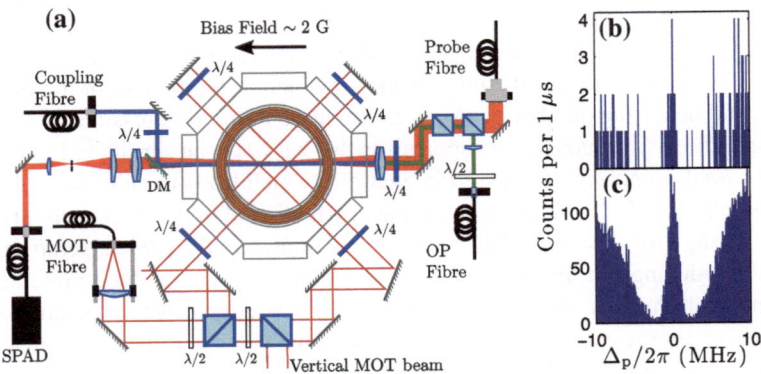

Fig. 7.10 **a** Modified setup to allow tight-focusing of probe beam down to $1/e^2$ radius of $12\,\mu$m, requiring a single-photon counter (SPAD) for detection. **b** Single shot EIT dataset recorded at 15 pW probe power **c** Histogram for 100 shot average clearly revealing the EIT resonance

used to spatially filter the optical pumping light, and the probe beam is coupled into a multi-mode fibre.

Using a tight probe focus means that the probe Rabi frequency is enhanced by a factor of $\sim$13, and so even for a 1 pW beam $\Omega_p/2\pi = 0.08$ MHz. At these weak probe powers, it is no longer possible to use a photodiode to detect probe transmission. Instead, a Perkin-Elmer SPCM-AQRH-15 single photon avalanche photodiode (SPAD) is used, chosen for its very low dark count of 42 counts/s. The SPAD output is not proportional to the probe intensity like a photodiode, instead a 15 ns TTL pulse is emitted when one or more photons is detected. The SPAD is therefore connected to a SensL HRMTime time correlated counting card which records the arrival times of these TTLs with 27 ps resolution. To protect the photon counter from damage due to a large photon flux, the detector is gated off using the circuit in appendix A.2 during the MOT load and optical pumping stage, and activated during the probe pulse. To avoid errors due to pile-up or saturation of the counter, the probe beam is attenuated after the chamber to give a count rate of 1 MHz, giving on average 1 count/μs. Data are recorded by taking 100 repeats to build up a histogram of arrival times, using a 1 μs bin width. The errors in each bin are assumed to be Poissonian, such that for each bin with m counts the standard deviation is $\sqrt{m}$ [25]. Transmission is then calculated from the histograms as described for the photodiode voltages in Sect. 6.5. An example of a histogram obtained in a single experimental run is shown in Fig. 7.10b for EIT on the $60S_{1/2}$ state with a 15 pW probe power. A single run gives a noisy outline of the EIT resonance, however after 100 repeats a much clearer lineshape is obtained, shown in (c). A final change from the old setup is an improvement in the stability of the coupling laser lock to reduce the ±300 kHz frequency jitter observed above.

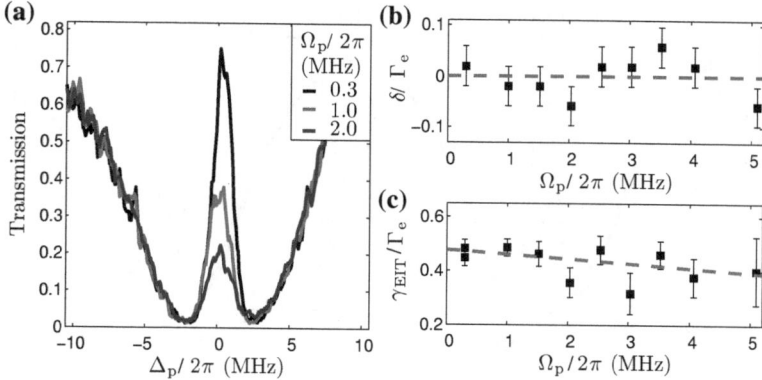

Fig. 7.11 a EIT spectra for $60S_{1/2}$ showing significant suppression of the resonant transmission with increasing probe Rabi frequency. **b** Fitting the detuning of the two-photon resonance δ and **c** FWHM linewidth γ_{EIT} show there is no shift or broadening associated with the suppression

7.4.2 Suppression Mechanisms

EIT spectroscopy is performed on the $60S_{1/2}$ state at a range of probe powers using a 1 s MOT load, which gives a density of $\rho = 1.2 \pm 0.1 \times 10^{10}$ cm^{-3} with around 7000 atoms contained within the probe volume. The results are shown in Fig. 7.11a which for the low Rabi frequency data shows a narrow EIT feature with 75 % transmission on the two-photon resonance, corresponding to a coupling Rabi frequency of $\Omega_c/2\pi = 4.6 \pm 0.1$ MHz. Another noteworthy feature is that the histogram represents data recorded over 100 s, however the effective linewidth of the two-photon resonance obtained from fitting is $\gamma_{rel}/2\pi = 110 \pm 50$ kHz, showing the coupling laser lock is much more stable than before. For increased probe powers, the data shows significant suppression of the resonant transmission by more than 50 % for $\Omega_p/2\pi = 2$ MHz whilst giving a completely symmetric EIT lineshape. From the spectra the detuning of the two-photon resonance, δ, and the FWHM of the EIT, γ_{EIT}, are determined, plotted in (b) and (c). These graphs show there is no shift or broadening of the EIT resonance accompanying the suppression, even for $\Omega_p/2\pi = 5$ MHz. It is now necessary to compare these observations to alternative mechanisms for suppression other than dipole blockade of the EIT dark state.

The lack of a shift in the data rules out ionisation as a suppression mechanism. This can be seen clearly from Fig. 7.12a which shows the results of the ion model described above calculated using experiment parameters, with the scalar polarisability of $\alpha_0 = 180$ MHz/(V/cm)^2 for the $60S_{1/2}$ state. Another comparison that can be made is to a mean-field model. As with the ion model, a random distribution of atoms is generated, and a fraction of these selected as Rydberg atoms. For the remaining atoms, the level shift is calculated by summing over the C_6/R^6 interaction energy with the Rydberg atoms. The resulting lineshape is found by summing over the susceptibility of each atom, using the interaction strength of $C_6 = -140$ GHz μm^6. The calculated spectra

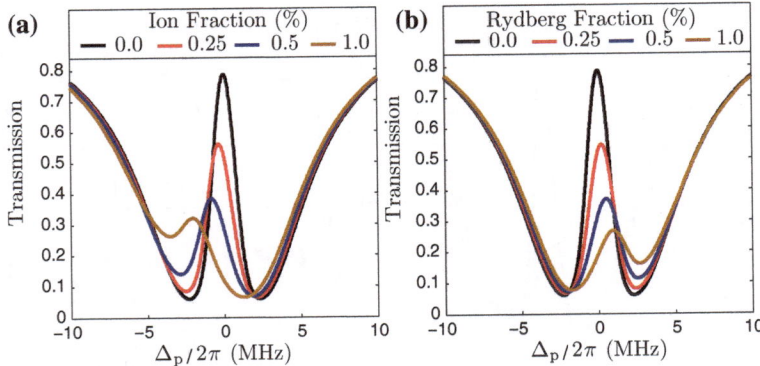

Fig. 7.12 Alternative suppression mechanisms. **a** Ion model showing the effect of Stark-shift, resulting in a large red-shift. **b** Mean-field model for the Rydberg interaction showing a blue-shift and broadening of the resonance accompanying the suppression

are plotted in Fig. 7.12b, which shows that whilst the mean-field model predicts suppression of the EIT resonance, this is associated with a shift and broadening of the two-photon resonance, as discussed in Sect. 5.4.3. This observation is important, as it validates the assumption that the mean-field treatment is incomplete and that the many-body cooperative model is an accurate description of the dynamics.

The final alternative suppression mechanism is van der Waals dephasing, which was introduced in Sect. 5.2. Van der Waals dephasing leads to an inhomogeneous broadening of the Rydberg states, caused by distribution of level shifts in the medium. To reproduce this effect, a single atom optical Bloch model was used in which the dephasing rate of the Rydberg state is proportional to the fraction of population in the Rydberg state, shown schematically in Fig. 7.13a. This is done by adding the dephasing rate $\gamma' = \gamma_{\text{vdW}}\sigma_{rr}$ to the relative laser linewidth terms γ_{rel} and γ_{c} in Eq. 4.14 to increase the dephasing rate of the coherence terms without changing the decay rate out of $|r\rangle$. The model is solved using the parameters obtained from the weak-probe fit for a range of Rabi frequencies and the value of γ_{vdW} optimised to reproduce the resonant transmission observed in the experiments. The results are shown in Fig. 7.13b–d for $\Omega_{\text{p}}/2\pi = 0.1, 2$ and $4\,\text{MHz}$ respectively for $\gamma_{\text{vdW}} \sim 7\,\Gamma_e$. These show that the dephasing can reproduce the resonant suppression for the 2 MHz data quite well, with only a slight broadening of the EIT resonance. For higher probe powers however, the EIT resonance is broadened significantly, and this also causes a broadening of the probe absorption which is not observed in experiment. The FWHM of the model traces is compared to the experiment in (e), clearly showing van der Waals dephasing cannot explain the observed suppression as there is no broadening in the data.

The result of this analysis is that the only mechanism consistent with observations is the cooperative optical non-linearity arising due to dipole–dipole interactions. The best proof for this however is to look for a non-linear density scaling to show the

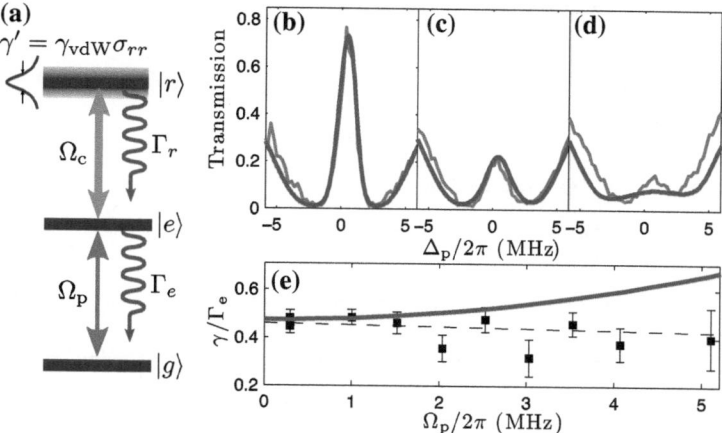

Fig. 7.13 Van der Waals dephasing model. **a** Schematic of model, with the dephasing rate of $|r\rangle$ proportional to the population in $|r\rangle$. **b–d** Comparison between theory and data for $\Omega_p/2\pi = 0.1$, 2 and 4 MHz. **e** FWHM for model compared to data, showing the dephasing model is not consistent with observations

optical response of a single atom is dependent upon interactions with the surrounding atoms.

7.4.3 Density Scaling

To test the density dependence of the suppression, transmission data is recorded at probe Rabi frequencies of $\Omega_p/2\pi = 0.1$ and 2.0 MHz as a function of density for the $60S_{1/2}$ and $54S_{1/2}$ states. Unlike the method described in Sect. 7.3.4, it is not possible to directly convert the transmission data into susceptibility using the relation $\chi_I = -\log_e(T)/k\ell$. This is because for the strong probe data, the transmission changes from 80 % at low density to 20 % at high density. Over this range, the non-linear absorption through the cloud cannot be neglected as the probe is attenuated as it propagates through the medium. The modification of the susceptibility due to the optical non-linearity will therefore vary strongly through the cloud at high density, and only very weakly at low density. This makes comparison between the susceptibility calculated in each regime unreliable. Instead, the optical depth $-\log_e(T_{\text{EIT}})$ is scaled relative to the probe-only optical depth $-\log_e(T_{\text{ABS}})$ to remove the first-order density dependence in the medium as shown in Fig. 7.14.

Looking first at the weak-probe data, this shows there is no non-linear density scaling for either state. This is expected from the weak probe dark state $|D\rangle = |g\rangle$, which means the optical response of each atom is independent of the surrounding atoms. For the strong probe however there is a very clear second-order density scaling,

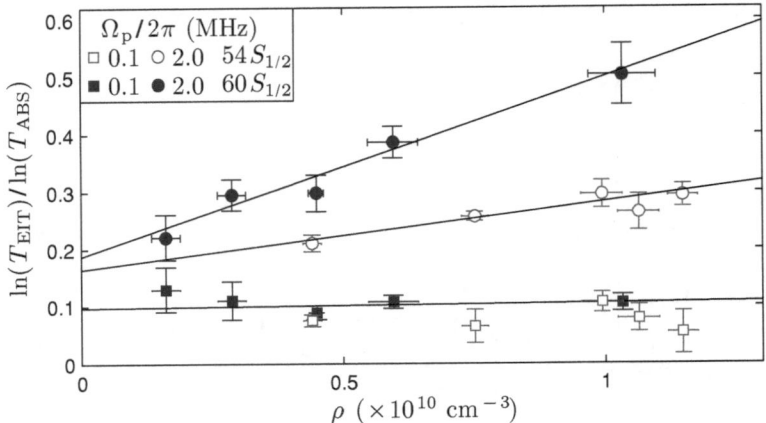

Fig. 7.14 Optical depth as a function of density, scaled relative to the probe-only optical depth to remove the trivial linear density scaling. This shows a second-order density dependence for the strong probe data, verifying the cooperative nature of the non-linearity

as seen above in Fig. 7.9 for the $D_{5/2}$ states, which verifies the cooperative nature of the EIT suppression.

The $60S_{1/2}$ state shows a steeper density dependence than is observed for $54S_{1/2}$. The difference in the gradients is due to the increased interaction strength of the $60S_{1/2}$ state compared to the $54S_{1/2}$ state, which leads to a larger blockade radius. Making the assumption $\gamma_{EIT} \propto \Omega_c$ (consistent with experiment observations) and using the scaling relations for $C_6 \propto n^{*11}$ and $\Omega_c \propto n^{*-3/2}$, the number of atoms in the blockade sphere should scale as $\mathcal{N}_b \propto R_b^3 \propto \sqrt{C_6/\gamma_{EIT}} \propto n^{*25/4}$. This gives a ratio of 2.0 for the number of atoms per blockade sphere for the two states. Applying a linear fit to the strong probe data, the ratio of the gradients is 2.7 ± 0.7 which is consistent with the suppression scaling with $\mathcal{N}_b$. The non-linearity can therefore be tuned by choice of density and principal quantum number, offering a high degree of control.

7.4.4 Comparison with the N-atom Model

Having proved the suppression is caused by the cooperative interaction between atoms, it is interesting to compare the experiment to the $\mathcal{N}$-atom model developed in Sect. 5.4. For the spectra presented in Fig. 7.11a the EIT linewidth is $\gamma_{EIT}/2\pi = 3\,\text{MHz}$, giving a blockade radius $R_b = 6\,\mu\text{m}$. At a density of $1.2 \times 10^{10}\,\text{cm}^{-3}$ this corresponds to an average of $\mathcal{N}_b = 11$ atoms per blockade sphere. Solving the model for this number of atoms is not possible due to the large number of states required, so instead the experiment is repeated at a density of $0.35 \pm 0.03 \times 10^{10}\,\text{cm}^{-3}$ which gives an average of $\mathcal{N}_b = 3$. Transmission data for $\Omega_p/2\pi = 0.1$ to $3.2\,\text{MHz}$ are shown in

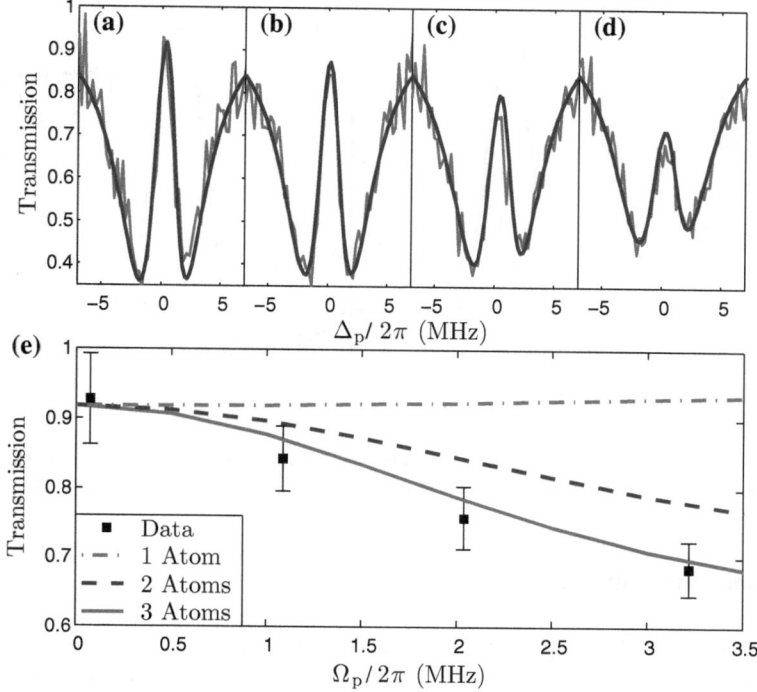

Fig. 7.15 Comparison of $\mathcal{N}$-atom model to data. **a–d** show EIT spectra recorded at a density of $0.35 \times 10^{10} \, \text{cm}^{-3}$ with an average $\mathcal{N}_b = 3$ for $\Omega_p/2\pi = 0.1, 1.0, 2.0$ and $3.2 \, \text{MHz}$ respectively. Transmission calculated using the three-atom model is plotted on top (*thick line*). **e** Resonant transmission compared to the $\mathcal{N}$-atom model for $\mathcal{N} = 1$–3

Fig. 7.14a–d, which shows the familiar suppression at increasing probe powers but by less than Fig. 7.11a due to the reduction in $\mathcal{N}_b$. Plotted on the data is the transmission calculated from the three-atom model using parameters of $\Omega_c/2\pi = 0.8 \, \text{MHz}$ and $\gamma_{\text{rel}} = 110 \, \text{kHz}$ obtained from the weak-probe fit to (a), changing only Ω_p between the figures. The only free parameter in the model is the interaction strength $V(R)$. The transmission spectra presented here are calculated using $V(R)/2\pi = 15 \, \text{MHz}$, however the result is insensitive to the interaction providing $V(R) > \gamma_{\text{EIT}}$ to match the blockade condition for the three atoms. The model is in excellent agreement with the data, reproducing not only the resonant transmission but also the full EIT lineshape. In (e) the resonant transmission is plotted as a function of Ω_p compared to the resonant transmission calculated for the 1, 2 and 3-atom models. This shows that the agreement for the three-atom model is better than for 2, as expected from the average $\mathcal{N}_b$ at this low density.

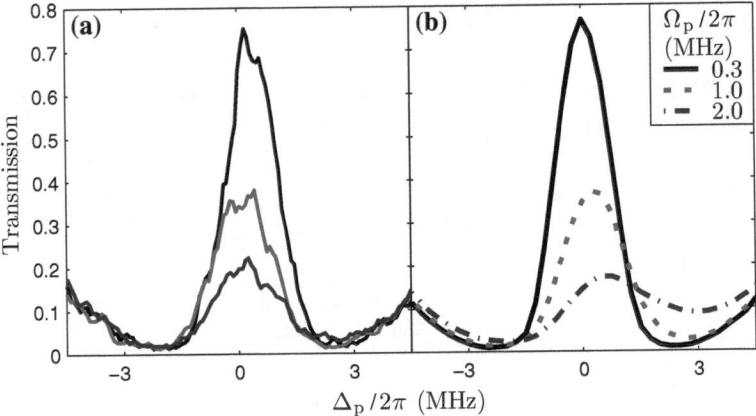

Fig. 7.16 Monte-Carlo model of the high density $60S_{1/2}$ EIT spectra. **a** Shows the experiment data and **b** the theoretical transmission. These calculations were performed by S. Sevinçli et al.

Blockade Dephasing Rate

The excellent quantitive agreement between the model and experiment show that the EIT is sensitive to the coherence of the blockaded ensemble, as only by considering the coherence of the many-body system is it possible to reproduce the observed suppression without broadening. Thus on the two-photon resonance the system evolves into an ensemble of blockaded ensembles at large probe power. Dephasing between neighbouring blockade spheres would lead to broadening of the EIT resonance, equivalent to an increase in the relative two-photon laser linewidth γ_{rel}. As there is no broadening observed in this regime (seen from Fig. 7.11c), this places an upper limit on the dephasing rate equal to the measured linewidth in the weak-probe regime. Thus the dephasing rate between neighbouring blockade spheres $< 110\,\mathrm{kHz}$ for the $60S_{1/2}$ state.

Monte-Carlo Modelling Results

Using the complete $\mathcal{N}$-atom model has shown excellent agreement with data for this low density data, however as it cannot be scaled to larger atom numbers its application is limited to this low $\mathcal{N}_b$ regime. As mentioned in Sect. 5.4, an alternative method has been developed by C. Ates, S. Sevinçli and T. Pohl [20] which uses a Monte-Carlo approach to model the steady-state populations for very large numbers of interacting atoms, from which the transmission can be calculated. As the EIT experiments are performed slow enough to be in the steady-state regime, this model has been used to reproduce the high density $60S_{1/2}$ data of Fig. 7.11a. The calculation makes no assumptions about how many atoms are in a blockade sphere, and instead creates a random distribution of atoms and calculates the real interaction strength with the surrounding atoms. It also fully accounts for the Gaussian distributions of the atomic density and laser intensity, making it more complete than the treatment presented above.

The results are presented in Fig. 7.15, with the data in (a) and the model output in (b). Looking first at the resonant transmission, the model very accurately reproduces the height of the EIT peak for each value of Ω_p. However, the calculated spectrum is asymmetric, showing evidence of both shift and broadening that is not observed in the experiment. The cause for the asymmetry in the model is the laser resonantly exciting the anti-blockaded pair states, which due to the repulsive dipole–dipole interactions lie at positive detunings. The reason these are not observed in the experiment is most likely due to motion from the strong van der Waals interactions. For a pair of atoms separated by the blockade radius of $6\,\mu m$, the interaction shift is $3\,MHz$. The laser takes a time of $38\,\mu s$ to scan from $\Delta_p/2\pi = 0$ to $3\,MHz$, during which time the atoms have been repelled to a separation around $10\,\mu m$ with an interaction energy of $0.5\,MHz$. Thus, by the time the laser reaches the resonance frequency of the short-range pair states, they have been repelled to long range. This argument is consistent with the slight asymmetry observed around $\Delta_p/2\pi = 2\,MHz$ for the low density data in Fig. 7.14d, as at low density the atoms are initially further apart, making the repulsive motion less important.

Complementary work has been done in the CPT regime discussed in Sect. 4.2.1, where the atoms are initially excited to the Rydberg state and the interacting CPT dark state studied by measuring the population remaining in $|r\rangle$ as a function of laser detuning [26]. Using this Monte-Carlo method, it has been shown that these two regimes can be collapsed onto a universal curve relating the transmission to the Rydberg population [20], and a more detailed comparison of this model to both experiments is presented in [27]. Unfortunately however, neither experiment measured both transmission and population. The universality therefore remains to be demonstrated.

7.4.5 Summary

Using the repuslive $S_{1/2}$ states it has been possible to conclusively demonstrate the observed suppression of the EIT for strong probe powers is due to the cooperative optical non-linearity arising from dipole–dipole interactions. This is characterised by a suppression without shift, which also displays a second-order density scaling dependent upon the number of atoms in the blockade volume, seen from the comparison of $54S_{1/2}$ to $60S_{1/2}$ in Fig. 7.16. Excellent agreement between experiment and the full $\mathcal{N}$-atom model is obtained at low density, reproducing both the resonant transmission and the full frequency spectrum. As EIT is a probe of the coherence between the atomic states, this allows an upper limit of $110\,kHz$ to be placed on the dephasing rate between neighbouring blockade spheres, showing the coherent interaction in each blockade volume is not destroyed by surrounding atoms at longer range.

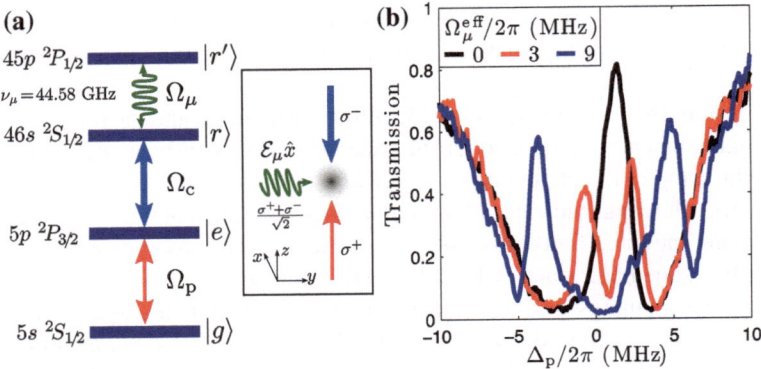

Fig. 7.17 Microwave Dressing **a** A linearly-polarised microwave field couples the $46S_{1/2}$ state to $45P_{1/2}$, which due to quantisation drives $\sigma^+ + \sigma^-$ transitions. **b** Microwave EIT showing Autler-Townes splitting of the resonance due to the microwave coupling

7.5 Microwave Dressing

The EIT experiments so far have all been performed in the regime of van der Waals interactions between the Rydberg states, with a $1/R^6$ interaction potential. An alternative is to resonantly couple two closely spaced Rydberg states together using a microwave field, as described in Sect. 3.4.2. This microwave dressing changes the interactions to resonant dipole–dipole which scale as $1/R^3$ resulting in a longer range interaction, and hence larger R_b.

To measure the effect of this change in interaction strength, the $46S_{1/2}$ state is resonantly coupled to the $45P_{1/2}$ state using microwaves at 44.58 GHz, as shown schematically in Fig. 7.17a, derived from an Anritsu MG3696A synthesiser. At this frequency it is necessary to use waveguide rather than coaxial cables, and a WR19 waveguide was used to direct the linearly polarised microwave field onto the cold atom cloud. EIT spectra taken for $\Omega_p/2\pi = 0.08$ MHz and $\Omega_c/2\pi = 5.5$ MHz are shown in Fig. 7.17b for different microwave Rabi frequencies, Ω_μ. Without the microwave field, a single EIT resonance is obtained. The effect of the strong microwave coupling is to create an Autler-Townes splitting of the Rydberg states, causing the EIT resonance to split proportional to $\propto \sqrt{\Omega_c^2 + \Omega_\mu^2}$.

As both the microwave polarisation and propagation direction were orthogonal to the quantisation axis along the probe axis, shown schematically in Fig. 7.17a, the microwave drives both σ^+ and σ^- transitions simultaneously. This makes the interpretation of the spectra more complex as the $\sigma^\pm$ transitions have different angular transition dipole matrix elements, giving a range of splittings for the Rydberg states. To reproduce the spectra it is necessary to use a 10-level model which includes all of the hyperfine levels of the two Rydberg states. This model was developed by M. Tanasittikosol and is described in detail in ref. [28]. Using this model, the effective

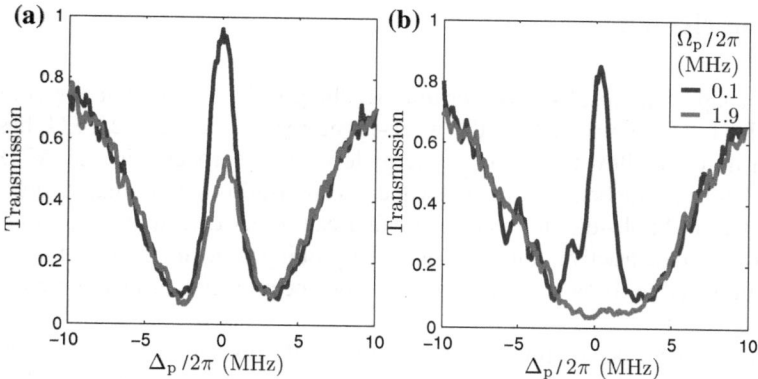

Fig. 7.18 Enhanced suppression from microwave dressing of $46S_{1/2}$. **a** EIT spectroscopy without microwave field. **b** Applying a weak microwave field $\Omega_\mu^{\mathrm{eff}}/2\pi \sim 500\,\mathrm{kHz}$ results in a dramatic enhancement of the EIT suppression

microwave Rabi frequency, $\Omega_\mu^{\mathrm{eff}}$ was extracted by averaging over all of the transition strengths, enabling calibration of the microwave coupling strength.

To observe an enhancement of the non-linearity, it is necessary to apply a weak microwave coupling with $\Omega_\mu < \Omega_c$ for all transitions to prevent an Autler-Townes splitting of the EIT resonance. Data are recorded at $\Omega_p/2\pi = 0.1$ and $1.9\,\mathrm{MHz}$ without the microwave field, shown in Fig. 7.18a at a density of $\rho = 1.2\pm0.1\times10^{10}\,\mathrm{cm}^{-3}$. This shows suppression of the EIT, as expected from the previous experiments. Repeating this with $\Omega_\mu^{\mathrm{eff}}/2\pi \sim 500\,\mathrm{kHz}$, shown in (b) for the same probe powers, the enhancement in the non-linearity due to the microwave source is dramatic - there is no evidence of loss or asymmetry, instead the EIT is completely suppressed.

The $46S_{1/2}$ state has a van der Waals interaction strength of $C_6 = -5.6\,\mathrm{GHz}\,\mu\mathrm{m}^6$, which corresponds to a blockade radius of $R_b = 3.8\,\mu\mathrm{m}$ for the fitted EIT linewidth of $\gamma_{\mathrm{EIT}}/2\pi = 2\,\mathrm{MHz}$. Thus in (a) there are an average of 3 atoms in each blockade sphere contributing to the EIT suppression. Applying the resonant microwave coupling to the $45P_{1/2}$ state, the interactions are now resonant dipole–dipole with a strength of $C_3 \simeq 0.8\,\mathrm{GHz}\,\mu\mathrm{m}^3$ which gives $R_b \simeq 7.4\,\mu\mathrm{m}$, almost twice the size of blockade in the undressed system. Consequently, the average number of atoms in each blockade volume increases to around 20 which is sufficient to suppress the EIT almost to the probe-only transmission. Microwave dressing therefore provides a method to significantly enhance the optical non-linearity by increasing the blockade radius without requiring excitation to very high n states.

7.6 Conclusion

In this chapter the results of performing Rydberg EIT on an interacting cold atom cloud have been presented. In the weak-probe regime $\Omega_p \ll \Omega_c$, EIT has been demonstrated to allow the Rydberg energy levels to be probed with sub-MHz resolution. This has the benefit that the Rydberg state itself is not populated, preventing shifts due to dipole–dipole interactions that can affect detection using ionisation. It is also non-destructive, allowing repeated probing over timescales of ms. EIT is therefore suitable for applications in precision spectroscopy or electrometry of the Rydberg states.

In the strong probe regime however, Rydberg EIT reveals a wide range of cooperative phenomena. For the low-n states this was manifested as a superradiant loss from the Rydberg state on the two photon resonance. This is seen from the scaling of the data with kR, and for $19D_{5/2}$ has been successfully modelled as a density-dependent loss. Similar behaviour is also observed for $44D_{5/2}$, with higher n states required to observe the effects of dipole–dipole level shifts dominating over the collective broadening.

For the repulsive S-states, suppression of the EIT resonance due to a cooperative optical non-linearity has been conclusively verified, reproducing not only the nonlinear density scaling but also obtaining a quantitative agreement to theory for low density data with an average of $\mathcal{N}_b = 3$ atoms per blockade sphere. As EIT probes the coherence of the blockaded system, this places an upper limit of 110 kHz on the dephasing rate of the blockade sphere, making it suitable for applications in quantum optics. These results are significant, representing the first observation of a novel cooperative optical non-linearity in an atomic system, mediated by the tuneable, long-range dipole–dipole interactions of the Rydberg states.

Similar suppression was observed for the attractively interacting $D_{5/2}$ states, which show different non-linear scalings in the susceptibility dependent upon the direction of the probe frequency sweep, suggesting motional effects from the pairwise interactions play a role in the observed spectra. For both scan directions the non-linearity has been characterised, which again shows a quadratic density dependence expected for a cooperative effect. Finally, a microwave coupling was shown to dramatically enhance the suppression by switching the interactions to resonant dipole–dipole, providing an additional control of the non-linear effect alongside choice of $n\ell$ state.

One of the limitations of the present experiments is the lack of information about the ion fraction accompanying the optical spectra. Combining these two detection methods could provide a complete understanding of the hysteresis observed in the D state data and assess the relative importance of ionisation, motion and blockade in the optical non-linearity. From measurements of both transmission and Rydberg population for a range of ratios of Ω_p to Ω_c, the universal scaling predicted by Ates et al. [20] for the cooperative non-linearity could be tested.

Another area not considered is working with the coupling laser detuned off-resonance to measure a dispersive non-linearity due to the interactions. This would allow characterisation of the regime in which the non-linearity could be used for providing controlled phase-shifts of optical fields, rather than the present regime in which the light is attenuated by the effect. To measure phase shift directly, an interferometer is required with the atoms in one arm [29]. This avoids the issues encountered in trying to extract the imaginary part of susceptibility from the transmission and having to use the Kramers-Kronig relations to calculate the dispersion.

As will be seen in the next chapter, the cooperative non-linearity offers a far more subtle effect which has not been considered so far, but will present itself as far more relevant to the objective of creating single-photon non-linearities.

References

1. A.K. Mohapatra, T.R. Jackson, C.S. Adams, Coherent optical detection of highly excited Rydberg states using electromagnetically induced transparency. Phys. Rev. Lett. **98**(11), 113003 (2007)
2. H. Kübler, J.P. Shaffer, T. Baluktsian, R. Löw, T. Pfau, Coherent excitation of Rydberg atoms in micrometre-sized atomic vapour cells. Nat. Photon. **4**, 112 (2010)
3. M.G. Bason, M. Tanasittikosol, A. Sargsyan, A.K. Mohapatra, D. Sarkisyan, R.M. Potvliege, C.S. Adams, Enhanced electric field sensitivity of rf-dressed Rydberg dark states. New J. Phys. **12**, 065015 (2010)
4. A.K. Mohapatra, M.G. Bason, B. Butscher, K.J. Weatherill, C.S. Adams, A giant electro-optic effect using polarizable dark states. Nat. Phys. **4**, 890 (2008)
5. I.I. Beterov, I.I. Ryabtsev, D.B. Tretyakov, V.M. Entin, Quasiclassical calculations of blackbody-radiation-induced depopulation rates and effective lifetimes of Rydberg nS, nP, and nD alkali-metal atoms with n $\leq$ 80. Phys. Rev. A **79**(5), 052504 (2009)
6. K.J. Weatherill, J.D. Pritchard, R.P. Abel, M.G. Bason, A.K. Mohapatra, C.S. Adams, Electromagnetically induced transparency of an interacting cold Rydberg ensemble. J. Phys. B **41**(20), 201002 (2008)
7. T.A. Johnson, E. Urban, T. Henage, L. Isenhower, D.D. Yavuz, T.G. Walker, M. Saffman, Rabi oscillations between Ground and Rydberg states with dipole–dipole atomic interactions. Phys. Rev. Lett. **100**(11), 113003 (2008)
8. A. Grabowski, R. Heidemann, R. Löw, J. Stuhler, T. Pfau, High resolution Rydberg spectroscopy of ultracold rubidium atoms. Fortschr. Phys. **54**(8–10), 765 (2006)
9. T. Vogt, M. Viteau, J. Zhao, A. Chotia, D. Comparat, P. Pillet, Dipole blockade at Förster resonances in high resolution laser excitation of Rydberg states of cesium atoms. Phys. Rev. Lett. **97**(8), 083003 (2006)
10. M. Reetz-Lamour, T. Amthor, J. Deiglmayr, M. Weidemüller, Rabi oscillations and excitation trapping in the coherent excitation of a mesoscopic frozen Rydberg gas. Phys. Rev. Lett. **100**(25), 253001 (2008)
11. B. Sanguinetti, H.O. Majeed, M.L. Jones, B.T.H. Varcoe, Precision measurements of quantum defects in the $nP_{3/2}$ Rydberg states of ^{85}Rb. J. Phys. B **42**(16), 165004 (2009)
12. A. Tauschinsky, R.M.T. Thijssen, S. Whitlock, H.B. van Linden, van den Heuvell, and R. J. C. Spreeuw., Spatially resolved excitation of Rydberg atoms and surface effects on an atom chip. Phys. Rev. A **81**, 063411 (2010)
13. P.J. Tanner, J. Han, E.S. Shuman, T.F. Gallagher, Many-body ionization in a frozen Rydberg gas. Phys. Rev. Lett. **100**(4), 043002 (2008)
14. J.O. Day, E. Brekke, T.G. Walker, Dynamics of low-density ultracold Rydberg gases. Phys. Rev. A **77**(5), 052712 (2008)
15. T. Wang, S.F. Yelin, R. Côté, E.E. Eyler, S.M. Farooqi, P.L. Gould, M. Koštrun, D. Tong, D. Vrinceanu, Superradiance in ultracold Rydberg gases. Phys. Rev. A **75**(3), 033802 (2007)

16. J.D. Pritchard, A. Gauguet, K.J. Weatherill, C.S. Adams, Optical non-linearity in a dynamical Rydberg gas. J. Phys. B **44**(18), 184019 (2011)

17. T. Amthor, M. Reetz-Lamour, S. Westermann, J. Denskat, M. Weidemüller, Mechanical effect of van der Waals interactions observed in real time in an ultracold Rydberg gas. Phys. Rev. Lett. **98**(2), 023004 (2007)

18. A. Chotia, M. Viteau, T. Vogt, D. Comparat, P. Pillet, Kinetic Monte Carlo modelling of dipole blockade in Rydberg excitation experiment. New J. Phys. **10**, 045031 (2008)

19. W. Li, P.J. Tanner, Y. Jamil, T.F. Gallagher, Ionization and plasma formation in high n cold Rydberg samples. Eur. Phys. J. D. **40**(1), 27 (2006)

20. C. Ates, S. Sevinçli, T. Pohl, Electromagnetically induced transparency in strongly interacting Rydberg gases. Phys. Rev. A **83**(4), 041802 (2011)

21. L.V. Hau, S.E. Harris, Z. Dutton, C.H. Behroozi, Light Speed reduction to 17 metres per second in an ultracold atomic gas. Nature **397**, 594 (1999)

22. R.W. Boyd, *Nonlinear Optics*, 3rd edn. (Academic Press, USA, 2008)

23. T.J. Carroll, K. Claringbould, A. Goodsell, M.J. Lim, M.W. Noel, Angular dependence of the dipole–dipole interaction in a nearly one-dimensional sample of Rydberg atoms. Phys. Rev. Lett. **93**(15), 153001 (2004)

24. J.D. Pritchard, D. Maxwell, A. Gauguet, K.J. Weatherill, M.P.A. Jones, C.S. Adams, Cooperative atom-light interaction in a blockaded Rydberg ensemble. Phys. Rev. Lett. **105**(19), 193603 (2010)

25. I. G. Hughes, T.P.A. Hase, *Measurements and their Uncertainties: A Practical Guide to Modern Error Analysis* (Oxford University Press, Oxford, 2010)

26. H. Schempp, G. Günter, C.S. Hofmann, C. Giese, S.D. Saliba, B.D. DePaola, T. Amthor, M. Weidemüller, S. Sevinçli, T. Pohl, Coherent population trapping with controlled interparticle interactions. Phys. Rev. Lett. **104**(17), 173602 (2010)

27. S. Sevinçli, C. Ates, T. Pohl, H. Schempp, C.S. Hofmann, G. Günter, T. Amthor, M. Weidemüller, J.D. Pritchard, D. Maxwell, A. Gauguet, K.J. Weatherill, M.P.A. Jones, C.S. Adams, Quantum interference in three-level Rydberg gases: coherent population trapping and electromagnetically induced transparency. J. Phys. B **44**, 184018 (2011)

28. M. Tanasittikosol, J.D. Pritchard, D. Maxwell, A. Gauguet, K.J. Weatherill, R.M. Potvliege, C.S. Adams, Microwave dressing of Rydberg dark states. J. Phys. B *44*(18), 184020 (2011). http://stacks.iop.org/0953-4075/44/i=18/a=184020

29. S.A. Aljunid, M.K. Tey, B. Chng, T. Liew, G. Maslennikov, V. Scarani, C. Kurtsiefer, Phase shift of a weak coherent beam induced by a single atom. Phys. Rev. Lett. **103**(15), 153601 (2009)

Part III
Rydberg Atom Quantum Optics

Chapter 8
Photon Blockade

In the preceeding chapters the laser field has been treated in the semi-classical approximation, where the electric field is represented by $E = E_0 \cos(\omega t)$. This assumption is adequate for considering the interaction between strong laser fields with macroscopic ensembles of independent atoms, as in this limit the quantum description of the light-field is indistinguishable from the classical treatment, for reasons that will be discussed below. However, in order to exploit the effect of the cooperative non-linearity at the single-photon level it is necessary to consider the quantised electromagnetic field, without which the concept of a photon becomes meaningless. Importantly, in quantum optics it is not the amplitude of the electric field, but rather the temporal and spatial correlations of the field that reveal the non-classical nature of light. Before considering the cooperative effect, it is necessary to first outline some fundamental ideas of the quantised field.

8.1 The Quantised Electric Field

A full derivation of the quantisation of the electromagnetic field can be found in many standard quantum optics textbooks e.g. [1], and here only the final results are detailed. The electric field is quantised in a finite volume V to obtain a set of spatial modes described by wavevector $\mathbf{k}$, each of which has two transverse polarisations λ defined by the polarisation unit vector, $\hat{e}_{\mathbf{k},\lambda}$. Each mode represents a quantum harmonic oscillator with a ladder of energies separated by $\hbar\omega_{\mathbf{k}}$, where $\omega_{\mathbf{k}} = c|\mathbf{k}|$. In this picture, a photon corresponds to a single excitation of the oscillator mode. Photons are added or removed from the mode using the creation ($\hat{a}_{\mathbf{k},\lambda}^{\dagger}$) and annihilation ($\hat{a}_{\mathbf{k},\lambda}$) operators which act on the wavefunction $|n_{\mathbf{k},\lambda}\rangle$ representing the number of photons in mode $\mathbf{k}$ as follows,

$$\hat{a}_{\mathbf{k},\lambda}|n_{\mathbf{k},\lambda}\rangle = \sqrt{n_{\mathbf{k},\lambda}}|n_{\mathbf{k},\lambda} - 1\rangle, \quad \hat{a}_{\mathbf{k},\lambda}^{\dagger}|n_{\mathbf{k},\lambda}\rangle = \sqrt{n_{\mathbf{k},\lambda} + 1}|n_{\mathbf{k},\lambda} + 1\rangle. \quad (8.1)$$

J. D. Pritchard, *Cooperative Optical Non-Linearity in a Blockaded Rydberg Ensemble*, Springer Theses, DOI: 10.1007/978-3-642-29712-0_8, © Springer-Verlag Berlin Heidelberg 2012

These combine to give the photon number operator, $\hat{n}_{\mathbf{k},\lambda} = \hat{a}^{\dagger}_{\mathbf{k},\lambda}\hat{a}_{\mathbf{k},\lambda}$. Eigenstates of this operator are known as Fock states, with exactly n photons in the mode.

As a consequence of the quantum nature of light, the electric field amplitude and phase can no longer be known simultaneously. This is because they are conjugate variables of the field, analogous to position and momentum, which are therefore constrained by the Heisenberg uncertainty principle.[1] Instead, the electric field is represented by the operator [1]

$$\hat{E}(\mathbf{r}, t) = i \sum_{\mathbf{k},\lambda} \sqrt{\frac{\hbar\omega_{\mathbf{k}}}{2\epsilon_0 V}} \hat{\mathbf{e}}_{\mathbf{k},\lambda}(\hat{a}_{\mathbf{k},\lambda}(t)e^{-i\omega_{\mathbf{k}}t+i\mathbf{k}\cdot\mathbf{r}} - \hat{a}^{\dagger}_{\mathbf{k},\lambda}(t)e^{i\omega_{\mathbf{k}}t-i\mathbf{k}\cdot\mathbf{r}}),$$

$$= \hat{E}^{(+)}(\mathbf{r}, t) + \hat{E}^{(-)}(\mathbf{r}, t), \tag{8.2}$$

where $\hat{\mathbf{e}}_{\mathbf{k},\lambda}$ is the polarisation unit vector. The operator is separated into the positive and negative frequency components such that $\hat{E}^{(+)}$ contains only annihilation operators and $\hat{E}^{(-)}$ contains creation operators, with $[\hat{E}^{(+)}]^{\dagger} = \hat{E}^{(-)}$.

8.1.1 Coherent States

The electric field emitted by a laser above threshold is described by a superposition of Fock states, known as a coherent state $|\alpha\rangle$ [1]. The coherent state is defined as

$$|\alpha\rangle = \sum_n \frac{\alpha^n}{\sqrt{n!}}e^{-|\alpha|^2/2}|n\rangle, \tag{8.3}$$

which is an eigenstate of the creation and annihilation operators with eigenvalues of

$$\hat{a}|\alpha\rangle = \alpha|\alpha\rangle, \hat{a}^{\dagger}|\alpha\rangle = \alpha^*|\alpha\rangle, \tag{8.4}$$

where α is the complex amplitude of the state. The probability of observing n photons in this coherent state is

$$P_{\alpha}(n) = |\langle n|\alpha\rangle|^2 = \frac{\alpha^{2n}}{n!}e^{-|\alpha|^2}, \tag{8.5}$$

which is a Poissonian distribution with a mean-photon number of $\bar{n} = |\alpha|^2$, and a fractional uncertainty $\Delta n/\bar{n} = 1/\sqrt{\bar{n}}$. The coherent states are minimum uncertainty states with equal uncertainty in phase and amplitude, thus for $\bar{n} \gg 1$ the coherent

---

[1] An important consequence of the uncertainty principle is spontaneous emission, which arises due to the coupling between an atom in the excited state and the vacuum fluctuations for the $|0\rangle$ state of each mode [2, 3].

state electric field $\langle\alpha|\hat{E}|\alpha\rangle = E_0\cos(\omega t)$ [1], equivalent to the semi-classical laser field used previously.

Having laid out a framework in which the photon can be defined, it is instructive to consider how to discriminate between a coherent state and non-classical light.

8.2 Photon Statistics

For a classical electric field, a photodiode can be used to generate a continuous signal proportional to the intensity of the field. Similarly, for a quantum field, the intensity at a detector is related to the expectation value for the intensity $\langle\hat{I}(\mathbf{r},t)\rangle$, where the intensity operator is defined as $\hat{I} = \hat{E}^{(-)}(\mathbf{r},t)\cdot\hat{E}^{(+)}(\mathbf{r},t)$. However, this is the same for both a single photon state and a coherent state with $\bar{n} = 1$. It is therefore insufficient to simply measure the field intensity, instead it is necessary to consider the photon statistics of the input field.

The photon statistics can be quantified using the second-order correlation function, also known as the intensity correlation function. For a pair of detectors at positions $\mathbf{r}_1$ and $\mathbf{r}_2$, the normalised second-order correlation function is defined as

$$g^{(2)}(\mathbf{r}_1,\mathbf{r}_2,t,t') = \frac{\langle\hat{E}^{(-)}(\mathbf{r}_1,t)\hat{E}^{(-)}(\mathbf{r}_2,t')\hat{E}^{(+)}(\mathbf{r}_2,t')\hat{E}^{(+)}(\mathbf{r}_1,t)\rangle}{\langle\hat{E}^{(-)}(\mathbf{r}_1,t)\hat{E}^{(+)}(\mathbf{r}_1,t)\rangle\langle\hat{E}^{(-)}(\mathbf{r}_2,t')\hat{E}^{(+)}(\mathbf{r}_2,t')\rangle}, \tag{8.6}$$

which describes the correlations between the field at time t and t'. If a continuous light source is used, the relative time t' can be related to a delay τ using $t' = t + \tau$, reducing this to the evaluation of $g^{(2)}(\tau)$.

For a classical field, the correlation function is bounded by the Cauchy–Schwarz inequality [4]

$$g^{(2)}(0) \geq 1, \tag{8.7}$$

however for a quantum field $0 \leq g^{(2)}(0) \leq \infty$. The $g^{(2)}$ function can therefore be used as evidence of a quantum or non-classical light field if $g^{(2)}(0) < 1$.

If the electric-field is single mode, the correlation function can be written in terms of creation and annihilation operators to simplify evaluation of the correlations,

$$g^{(2)}(\tau) = \frac{\langle a^\dagger(t)a^\dagger(t+\tau)a(t+\tau)a(t)\rangle}{\langle a^\dagger(t)a(t)\rangle\langle a^\dagger(t+\tau)a(t+\tau)\rangle}. \tag{8.8}$$

For a coherent state, $g^{(2)}(\tau) = 1$ for all time independent of α, as expected for a classical plane wave field. For a Fock state $|n\rangle$, $g^{(2)}(0) = 1 - 1/n$. Thus for the single photon state $g^{(2)}(0) = 0$, violating the classical inequality as expected for this purely quantum state of light. The physical interpretation of this result is that as there is only a single photon, it cannot be simultaneously observed by both detectors. This effect is known as *anti-bunching*, applicable to all states with $g^{(2)}(0) < 1$, as photons arrive at well spaced intervals compared to the random distribution of arrival times

Fig. 8.1 Hanbury Brown
Twiss interferometer. The
input light is separated onto a
pair of single photon counters
using a 50:50 beam-splitter,
and coincidence counts are
detected as a function of delay
τ to build up the correlation
function

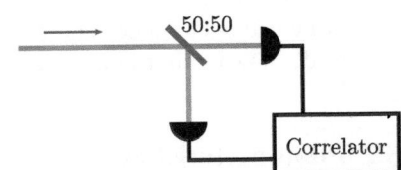

expected for the coherent state. Similarly, states with $g^{(2)}(0) > 1$ are *bunched* as
photons are more likely to arrive together.

In practise, $g^{(2)}(\tau)$ is measured with a Hanbury Brown Twiss (HBT) interferometer [5], shown schematically in Fig. 8.1 which uses a 50:50 beam-splitter to
separate light onto a pair of photon counters. The correlator records the coincidences
between the counters as a function of delay τ which can be used to determine the
normalised correlation function. The first experimental evidence of anti-bunching
was from observation of suppressed correlations at $\tau = 0$ in the resonance fluorescence of a single sodium atom [6], followed by the measurement of $g^{(2)}(0) = 0$ for
fluorescence of a single ion [7].

8.3 Photon Blockade

As every mode of the quantised electric field is a harmonic oscillator, there is a discrete
ladder of energies which for laser light is initially populated with the Poissonian
distribution of the coherent state. However, if the harmonicity of this ladder can be
broken, it is possible to observe non-classical states of light. An example of this is the
interaction between a single atom and the mode inside an optical cavity. The effect of
the atom-light interaction is to create an anharmonic energy ladder dependent upon
the photon occupation, as illustrated in Fig. 8.2a. The interaction causes the cavity
to be shifted off resonance with the probe laser after the first photon is absorbed,
preventing another photon from entering the cavity until the first photon leaves.
This system allows a coherent state to be filtered into a train of single photons, an
effect known as a *photon blockade* [8] or a *photon turnstile*. The resulting anti-
bunched output has been observed experimentally for optical cavities [9, 10], and
more recently in a superconducting microwave cavity [11], where an artificial atom
is used to overcome the limited fidelity in optical cavities due to residual motion of
the atom.

For the cooperative non-linearity due to dipole blockade, a similar effect can be
realised. In Chap. 5, the suppression of transmission was interpreted as the formation
of an entangled state with one atom in the EIT dark state and the remaining atoms
resonantly scattering light from the probe beam. Combining this with the concept
of a quantised electric field, the formation of the single collective dark state can
only involve a single photon from the probe field. Any other photons arriving in the
blockaded ensemble now resonantly couple to the excited state of the atoms and are

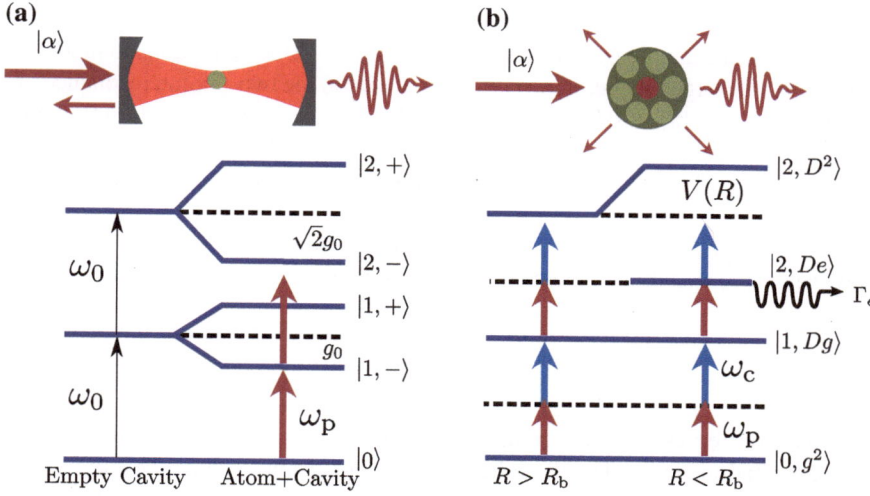

Fig. 8.2 Photon blockade. **a** Placing an atom at the centre of an optical cavity causes the cavity modes to be detuned by $\pm\sqrt{n}g_0$, where g_0 is the coupling constant, preventing more than a single photon entering the cavity. (States are labelled $|n, \pm\rangle$ to denote the photon number n and dressed state of the atom) **b** For the EIT system with no interactions, all photons form the dark state $|D\rangle$, so for two photons there are two dark states $|2, D^2\rangle$. Dipole–dipole interactions detune this state, breaking the EIT condition for the second photon and causing it to couple to the intermediate excited state $|2, De\rangle$. This state decays at rate Γ_e, scattering the photon into a different mode so only a single photon remains in the probe beam

scattered out of the mode of the probe beam, as shown schematically in Fig. 8.2b. This allows only a single photon to pass through un-scattered, resulting in a single photon output in the mode of the probe laser.

An important difference between these two schemes is that for the atom+cavity system, the energy of the optical transition is shifted for $n > 1$ so the cavity completely rejects all but a single photon, ensuring there will never be multiple photons at the cavity output. The Rydberg blockade mechanism, however, does not shift the energy of the optical transition for the probe laser. Instead, the medium changes from being transparent for the first photon, to opaque for $n > 1$. The photon scattering process for $n > 1$ is probabilistic, which may limit the fidelity as a photon turnstile. Nonetheless, it provides a mechanism to generate non-classical states of light from an input field initially in a coherent state without the need for an optical cavity.

8.3.1 Dark State Polariton

To aide the interpretation of a single-photon dark state, it is useful to introduce the concept of a *dark state polariton* [12]. In the analysis of EIT in Sect. 4.2, the dressed states of the atom were introduced to explain the EIT dark state assuming a constant amplitude driving field. For a weak probe beam, the probe electric field can be coupled to the atomic evolution using the Maxwell-Bloch equations to model the propagation through the medium. The result is that on the two-photon resonance, the system forms a stable, lossless quasi-particle known as a dark state polariton Ψ [13]

$$\Psi(z, t) = \cos \vartheta \mathcal{E}_{\mathrm{p}}(z, t) - \sin \vartheta \sqrt{\rho} \sigma_{g,r}(z, t) e^{i\Delta k z}, \qquad (8.9)$$

where $\mathcal{E}_{\mathrm{p}} = E_{\mathrm{p}}/\sqrt{\hbar\omega_{\mathrm{p}}/2\epsilon_0}$ is the normalised probe amplitude, $\Delta k = k_c - k_{\mathrm{p}}$ is the wave-vector mismatch, and the mixing angle ϑ is related to the group index,

$$\tan^2 \vartheta = \frac{6\pi c\rho \Gamma_e}{k_{\mathrm{p}}^2 \Omega_c^2} = \mathrm{n_{gr}}. \qquad (8.10)$$

The polariton represents a coherent superposition of the electromagnetic field and atomic excitation, denoted by the coherence σ_{rg}. For a small group index, the mixing angle is small and the electromagnetic component of the polariton dominates, with a group velocity around c. For a large group index however, the energy from the probe field is transferred to the atomic excitation, giving the field 'mass' and enabling slow propagation at speed $v_{\mathrm{gr}} \ll c$. At the edge of the medium, the excitation is converted back into an electromagnetic field without loss due to the perfect transmission achieved in EIT. Treatment of the probe as a quantised field yields equivalent results, with the electric field replaced by the electric field operator from Eq. 8.2 [12].

The polariton picture gives two insights relevant to achieving photon blockade. The first is that it shows that a large group index is required in order to transfer the single-photon field into atomic excitation in the medium. Without this, the Rydberg state is not populated and there is no dipole blockade. The second follows on from this, as the requirement of a large group-velocity means the single-photon polariton propagates slowly through the medium. During this propagation time, subsequent photons entering the medium should be scattered, introducing a characteristic delay between photon emission at the output which will be referred to as the blockade time, τ_{b}. In the limit of a strong driving field, this should result in a regular train of single photons separated by time τ_{b}.

Applying the condition for ϑ to the experiments presented in Fig. 7.11a for the suppression of transmission for the $60S_{1/2}$ state, the weak-probe group index was $\mathrm{n_{gr}} \sim 4 \times 10^4$, corresponding to a mixing angle of $\sim 90°$. For these experiments, the polariton is almost entirely composed of atomic excitation, meaning these photon-statistics must play a role in the observed suppression. However, comparing the blockade radius of $R_{\mathrm{b}} \sim 5 \, \mu\mathrm{m}$ to the $10 \, \mu\mathrm{m}$ $1/e^2$ radius of the beam waist shows the probe laser interacts with of order 16 blockade spheres over the beam cross-section. Thus, whilst each blockade region can potentially create a single-photon, the total output can still have as many as 16 photons which makes the direct observation of non-classical light in the probe beam challenging using $g^{(2)}$, as will be demonstrated below.

This analysis of the photon blockade due to dipole–dipole interactions gives a qualitative description of the mechanism, but a more quantitative approach is required to determine the potential fidelity and parameter range in which photon blockade can be realised. Solving the complete quantum dynamics for a quantised field coupled to an interacting $\mathcal{N}$-atom system is a non-trivial problem, and will not be attempted here. However, it is still possible to gain an insight into the expected correlations for the light output from a single blockaded ensemble. This will be the subject of the following sections.

8.4 Simple Model for $g^{(2)}(\tau)$

As a first attempt at predicting the photon statistics of the probe beam output from the blockade region, a simple model of the blockade mechanism is developed. Consider an ensemble of $\mathcal{N}$-atoms confined within a sphere of diameter R_b to ensure all atoms meet the blockade condition. This is probed by a tightly focused laser beam with a $1/e^2$ radius of $w_0 < R_b/2$ such that the probe beam is completely contained within the interaction volume to enable complete absorption of the probe beam. As mentioned above, the probe laser can be represented as a coherent state $|\alpha\rangle$ with a Poissonian distribution of photon numbers, however it is necessary to determine the mean photon number for $|\alpha\rangle$. To do this, a quantisation volume must be defined, which is trivial for a cavity but not for light in free-space. The purpose of the model is to determine the coincidences of photon arrival times at a detector, so quantisation can be achieved by defining a time window Δt in which photons are binned. In this time light travels a distance $c\Delta t$, so the probe beam can be quantised by introducing a cylindrical volume $V = \pi w_0^2 c \Delta t$ as illustrated in Fig. 8.3a, where the cylinder is assumed to have a radius equal to the beam waist. For a probe of power P, the mean photon number can then be calculated using Eq. C.5,

$$\bar{n} = \frac{2P\Delta t}{\hbar\omega}. \tag{8.11}$$

From the mean photon number, a random input photon train $C(t)$ is generated for 10^6 time-bins of width Δt, with the photon number in each time bin determined from the Poissonian distribution of Eq. 8.5. To model the effect of the photon blockade, two output modes are defined—a forward channel $C_f(t)$, which represents the probe light on the other side of the atomic ensemble, and a scattered mode $C_s(t)$, which represents all other modes in which photons can be scattered by the interaction with the medium. Starting at $t = 0$, the first photon to arrive in the medium is placed in the forward channel and all photons arriving within a period of τ_b are put in the scattered channel. This process is repeated across the entire photon train, as illustrated schematically in Fig. 8.3b.

During the propagation of the slow polariton, the other photons are scattered by resonantly coupling on the two-level transition between $|g\rangle$ and $|e\rangle$. If the ensemble is not completely optically thick on this probe-only transition, then the efficiency of this scattering process is limited. To account for this effect, photons in the scattered

(a)   **(b)**

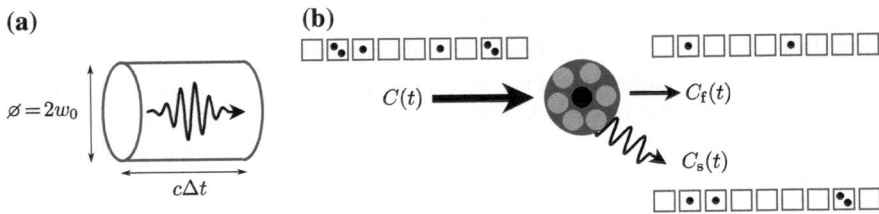

Fig. 8.3 Simple $g^{(2)}$ model. **a** The probe laser field is quantised in a cylinder of length $c\Delta t$. The probe power can then be converted to $\bar{n}$ to randomly generate photons in each window Δt. **b** The blockade mechanism is simulated by dividing the input photon train $C(t)$ into the forward mode of the probe beam $C_f(t)$, with only one photon passing through in a period τ_b, and a scattered channel $C_s(t)$ for the remaining photons. Each *box* represents a time period Δt, with *dots* showing photon number in each window

channel can be transferred back into forward channel with a probability equivalent to the probe-only transmission, which is parameterised in terms of the optical depth $OD = -\log_e(T)$.

Finally, the second order correlation function is calculated for each of the two output modes. This is achieved by taking the photon train $C_i(t)$ and simulating the effect of a beam-splitter to separate it into the counts detected by a pair of detectors D_1 and D_2, equivalent to the HBT interferometer in Fig. 8.1. If there are n photons in a given time-bin, the probability of detecting m photons at detector D_1 can be found using the binomial distribution

$$P(m) = \frac{n!}{m!(n-m)!} p^m (1-p)^{(n-m)}, \tag{8.12}$$

where p is the probability of success which for a 50:50 beam-splitter is 0.5. This distribution allows the beam-splitter to be modelled efficiently to obtain the counts arriving at the first detector in each time window, $D_1(t)$, from which $D_2(t) = C_i(t) - D_1(t)$. The normalised correlation function is then calculated using the Weiner-Khintchine theorem [14] as

$$g^{(2)}(\tau) = \text{Re} \left\{ \frac{\mathcal{F}^{-1}[\mathcal{F}[D_1(t)]\mathcal{F}[D_2(t)]^*]}{\sum D_1(t) \sum D_2(t)} \right\}, \tag{8.13}$$

where $\mathcal{F}$ and $\mathcal{F}^{-1}$ denote the Fourier transform and its inverse.

Having introduced the $g^{(2)}$ model for the correlations, it is useful to explicitly define the optical depth in terms of physical parameters. From the definition of transmission in Eq. 4.23a, $OD = -\log_e(T) = k_p \ell \chi_I$. Taking $\ell = R_b$ and using the weak-probe limit for the probe-only susceptibility χ_I from Eq. 4.20, the optical depth is given by

$$OD = k_p \cdot \frac{2\rho d_{eg}^2}{\epsilon_0 \hbar \Gamma_e} \cdot R_b = \frac{6\pi R_b \rho}{k_p^2} = \frac{36 \mathcal{N}}{k_p^2 R_b^2}, \tag{8.14}$$

where Eq. 5.3 has been used to eliminate d_{eg}^2 and a uniform density approximation used to obtain $\rho = 6\mathcal{N}/\pi R_b^3$. It is then trivial to re-scale the group index, and hence velocity, of the dark state polariton from Eq. 8.10 in terms of OD as

$$n_{gr} = \frac{\text{OD}c\,\Gamma_e}{R_b\,\Omega_c^2}, \quad v_{gr} = \frac{c}{1+n_{gr}} \simeq \frac{R_b\,\Omega_c^2}{\text{OD}\,\Gamma_e}. \tag{8.15}$$

This results in the following simple relation for the blockade time,

$$\tau_b = \frac{R_b}{v_{gr}} = \frac{\text{OD}\,\Gamma_e}{\Omega_c^2}. \tag{8.16}$$

Combining these relations together, we consider the case of $\mathcal{N} = 200$ confined within a blockade radius of $R_b \sim 5\,\mu$m, corresponding to a density of $\rho \sim 3 \times 10^{12}\,\text{cm}^{-3}$. This is two orders of magnitude larger than the MOT density, however this is achievable using an optical dipole trap, as will be discussed in Sect. 9.2. The optical depth for this case is OD $= 4.4$, resulting in 99% probability for scattering photons out of the probe beam. Taking $\Omega_c = \Gamma_e$ (consistent with a $5\,\mu$m blockade radius for $60S_{1/2}$) the corresponding blockade time is $\tau_b = 120\,$ns. The probe laser is assumed to be focused to a waist of $w_0 = 1\,\mu$m to satisfy $w_0 < R_b/2$, and the model is run for probe powers of $500\,$fW and $10\,$pW, equivalent to $\Omega_p/2\pi = 0.6$ and $2.6\,$MHz respectively.

The results for low power are shown in Fig. 8.4a which shows significant anti-bunching up to $\tau = \tau_b$ in the forward mode, and bunching for the scattered mode. This occurs because at low power there is a very low probability of observing any photons, so a large proportion of the photons arrive in the medium separated by times $t > \tau_b$ and pass through. For the scattered channel, there are now a relatively large fraction of multi-photon events compared to Poissonian statistics as the single photon component is suppressed, giving the observed bunching. For the strong probe results in (b), the bunching of the scattered channel becomes insignificant as most photons are scattered, with only a very slight change in the photon count distribution from the Poissonian input. In the forward channel however, a periodic anti-bunching is observed with strong bunched peaks at harmonics of $\tau = \tau_b$. These spikes are asymmetric as it is not possible for photons to arrive closer in time than τ_b, but the next photon may arrive at anytime later, smearing out the sharp peak. This also damps the amplitude of peaks at later times. For higher powers, the probability of having at least one photon in each time step Δt tends to unity, causing the $g^{(2)}$ to look more like a comb of delta-functions.[2]

To explore the dependence on the optical depth, the model is run for a $10\,$pW probe with $\mathcal{N} = 100$ and 400, corresponding to OD $= 2.2$ and 8.8 respectively. The results are shown in (c) compared to $\mathcal{N} = 200$, with the variation in τ_b clearly

---

[2] This sharp-edged correlation function is similar to that predicted for a $p - i - n$ junction in which the Coulomb blockade prevents more than a single photon emission [15], however the output coupling efficiency for such devices is too weak to measure the correlations [16].

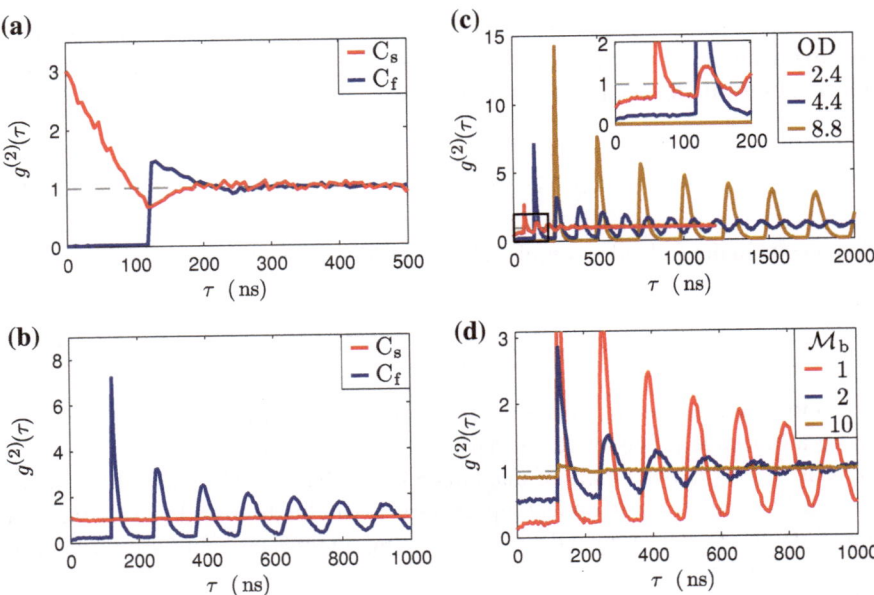

Fig. 8.4 $g^{(2)}$ Model results. Correlation function for OD $= 4.4$ calculated for **a** $P = 500\,\text{fW}$ and **b** $10\,\text{pW}$, clearly showing a strong anti-bunching for $\tau < \tau_b$. **c** Effect of varying OD for the forward mode C_f at $10\,\text{pW}$, showing at OD ~ 2 the anti-bunching is heavily suppressed. **d** Changing the number of blockaded ensembles $\mathcal{M}_b$ in the probe beam for C_f at $10\,\text{pW}$, OD $= 4.4$ shows even two blockade regions significantly reduces the visibility of the anti-bunching

visible from the arrival of the first peak. In the inset, the effect of small optical depth is easy to see, as it suppresses the anti-bunching at short times and rapidly damps out the peak visibility. An optical depth of OD $\gtrsim 4$ is therefore required to observe blockade experimentally.

Another effect that can be added to the model is having more than one blockaded volume in the cross-section of the beam. This is achieved by randomly splitting the input train $C(t)$ between $\mathcal{M}_b$ blockade regions, and performing the scattering on each blockade independently. The forward scattering from each is then combined, and the correlation of the total output found. Results calculated for the original parameters of OD $= 4.4$ are shown in (d). As more blockade regions are included, the anti-bunching of the output light is suppressed as $\mathcal{M}_b$ photons can propagate through the medium, which for $\mathcal{M}_b \gg \bar{n}$ allows the initial coherent state to be unchanged. Thus the visibility of the anti-bunching of the output light is very small for the experiments of Sect. 7.4 with $\mathcal{M}_b = 16$, as mentioned above.

In summary, these results show that the blockaded ensemble can be used to create a regularly spaced, highly correlated train of single photons, analogous to creating 'hard-edge' photons in a 1D lattice. The repetition rate of the photon pulses is $\tau_b^{-1} \sim$ MHz, which could be used as a semicontinuous single-photon source for quantum information. The fidelity of the single photon output state is limited by the optical depth of the ensemble, however for 400 atoms the model predicts $g^{(2)}(\tau < \tau_b) < 10^{-2}$ which is smaller than the uncertainty in a typical measurement of $g^{(2)}$ [9, 10]. An implicit assumption of this simple model is that the polariton is formed as soon as the photon is in the medium. However, there may be a finite timescale associated with the formation of a polariton. During this time two-photons could pass through the medium, which would compromise the fidelity. These EIT transients are considered in the next section.

From this simple model, it has been possible to verify the parameter range over which photon blockade can be realised, requiring an optical depth equivalent to several hundred atoms confined within a single blockade volume. This clearly represents a complex system to model rigorously, however if we consider the case of only a few atoms it is possible to calculate the correlations of the scattered field.

8.5 Resonance Fluorescence Correlation Functions

8.5.1 The Source-Field Expression

In Chap. 5 an $\mathcal{N}$-atom model was developed to calculate the properties of the interacting EIT system. Whilst this model is based on classical driving fields, these optical Bloch equations can be used to calculate the properties of the scattered light field from the atoms using the source-field expression [1]. This states that the electric field operator at position $\mathbf{r}$ is given by $\hat{\boldsymbol{E}}^{(+)}(\mathbf{r}, t) = \hat{\boldsymbol{E}}_f^{(+)}(\mathbf{r}, t) + \hat{\boldsymbol{E}}_{sf}^{(+)}(\mathbf{r}, t)$, where $\hat{\boldsymbol{E}}_f^{(+)}(\mathbf{r}, t)$ is the incident field and $\hat{\boldsymbol{E}}_{sf}^{(+)}(\mathbf{r}, t)$ is the radiation field of the atomic dipole, known as the source-field term. This is the quantum analogue of the classical Ewald-Oseen extinction theorem [17], which describes 'absorption' as a destructive interference between the incident plane wave and the radiated dipole field.

For an ensemble of $\mathcal{N}$-atoms located at positions $\mathbf{r}_i$, the source-field term in the far field ($k|\mathbf{r} - \mathbf{r}_i| \gg 1$ for all i) is given by [4, 18][3]

$$\hat{\boldsymbol{E}}_{sf}^{(+)}(\mathbf{r}, t) = -\frac{k^2 (\mathbf{d}_{eg} \times \hat{\mathbf{r}}) \times \hat{\mathbf{r}}}{4\pi\varepsilon_0 r} \sum_i^{\mathcal{N}} e^{-ik\hat{\mathbf{r}} \cdot \mathbf{r}_i} \hat{\pi}_i^- (t - r/c), \qquad (8.17)$$

which is equivalent to the classical dipole radiation field of Eq. 5.2 with the dipole moment replaced with operator $d_{eg}\hat{\pi}^-$.

The source-field expression therefore relates the scattered electric field to the properties of the atomic system. If we consider only positions off-axis with respect to the probe and coupling lasers, the incident field $\hat{\boldsymbol{E}}_f^{(+)}(\mathbf{r}, t)$ vanishes, and the electric field reduces to a sum over the dipole operators for the system. Absorbing

---

[3] In [4] the atom is assumed to be at the origin, however for a finite displacement an additional phase-factor is required which can be found in Eq. 7.13 of [18].

the geometric factors into the function $f(\mathbf{r})$, the scattered electric field is $\hat{\mathbf{E}}^{(\pm)}(\mathbf{r}, t) =$ $f(\mathbf{r})\hat{\Pi}^{\mp}(\mathbf{r}, t - r/c)$, where $\hat{\Pi}^{\pm}$ are the combined raising and lowering operators for the system,

$$\hat{\Pi}^{\pm}(\mathbf{r}, t) = \sum_i^{\mathcal{N}} e^{\pm i k \hat{\mathbf{r}} \cdot \mathbf{r}_i} \hat{\pi}_i^{\pm}(t). \tag{8.18}$$

8.5.2 Correlation Function

Using this definition of the electric field, the second order correlation function of Eq. 8.6 can be written as

$$g^{(2)}(\tau) = \frac{\langle \hat{\Pi}^+(t)\hat{\Pi}^+(t+\tau)\hat{\Pi}^-(t+\tau)\hat{\Pi}^-(t)\rangle}{\langle \hat{\Pi}^+(t)\hat{\Pi}^-(t)\rangle\langle \hat{\Pi}^+(t+\tau)\hat{\Pi}^-(t+\tau)\rangle}, \tag{8.19}$$

where $\langle \ldots \rangle$ denotes a trace over the density matrix for the atomic system. The correlation function is calculated using the quantum regression theorem which gives [19, 20]

$$G(t; t') = \langle \hat{A}(t')\hat{B}(t)\rangle = \mathrm{Tr}\{\hat{A}\sigma_{\mathrm{cond}}(t; t')\} \qquad (t' \geq t) \tag{8.20}$$

where the σ_{cond} is the conditional density matrix defined at time t as $\sigma_{\mathrm{cond}}(t; t) = \hat{B}\sigma(t)$, which represents the state of the system after the action of $\hat{B}$ is applied.

Applying this theorem to Eq. 8.19 allows the steady-state density matrix σ^{ss} for the $\mathcal{N}$-atom system to be calculated from the optical Bloch equations derived in Sect. 5.4. The conditional density matrix is evaluated using

$$\sigma_{\mathrm{cond}}(0) = \hat{\Pi}^- \sigma^{\mathrm{ss}} \hat{\Pi}^+, \tag{8.21}$$

which describes the state of the system after a photon has been emitted. The conditional density matrix is then re-normalised and used as an initial condition for the same optical Bloch equations, which are integrated until time τ to obtain $\sigma_{\mathrm{cond}}(\tau)$. Finally, the second-order correlation function is

$$g^{(2)}(\tau) = \frac{\mathrm{Tr}\{\hat{\Pi}^- \sigma_{\mathrm{cond}}(\tau)\hat{\Pi}^+\}}{\mathrm{Tr}\{\sigma_{\mathrm{cond}}(0)\}}. \tag{8.22}$$

For large τ, the conditional density matrix will evolve back to the steady-state σ^{ss}, resulting in $g^{(2)}(\tau \gg 1) = 1$ as required. The fluorescence correlations therefore arise from the dynamic evolution of the system back to the steady-state after emitting the first photon.

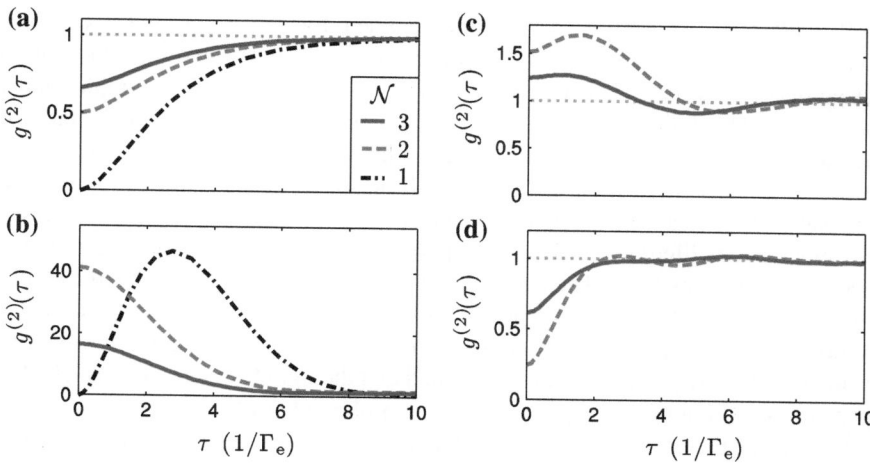

Fig. 8.5 $\mathcal{N}$-atom fluorescence correlations. **a** Independent two-level atoms for $\Omega_p = \Gamma_e/5$ displaying anti-bunching. **b–d** Interacting EIT system with $\Omega_c = \Gamma_e$, $V(R_{ij}) = 2\Gamma_e$ and $\Omega_p = \Gamma_e/5$, $\Gamma_e/2$ and Γ_e respectively shows blockade causes bunching, which becomes anti-bunched if the strong probe violates the blockade condition

8.5.3 Cooperative Emission from Incoherent Atoms

As a first approximation, the atoms are assumed to be incoherent emitters such that the cross-phase factors average to zero, for example due to atomic motion. In this case, the combined operators become separable to give

$$\hat{\Pi}^- \sigma \hat{\Pi}^+ = \sum_i^{\mathcal{N}} \hat{\pi}_i^- \sigma \hat{\pi}_i^+. \tag{8.23}$$

The resonance fluorescence correlations of independent two-level atoms for $\Omega_p = \Gamma_e/5$ is shown in Fig. 8.5a, showing anti-bunching with $g^{(2)}(0) = 1 - 1/\mathcal{N}$ as each atom can emit a single photon at a random time, with the possibility to observe two photons at zero delay from two atoms but with a non-Poissonian probability. In (b) the correlation function for the EIT system with $\Omega_c = \Gamma_e$ and $V(R_{ij}) = 2\Gamma_e$ is plotted. The $\mathcal{N} = 1$ trace is anti-bunched at $\tau = 0$, and then increases to give $g^{(2)}(\tau) \gg 1$ at $\tau \sim 1/\Omega_p$. This occurs because in the resonant EIT condition, the emission of a photon projects the atom out of the dark state, requiring another photon to be emitted at a later time to allow the atom to return to the dark state. Assuming a perfect laser system, this would not be observable in an experiment as the probability to emit the initial photon is vanishing due to the EIT condition. This very small probability for emission of the first photon leads to an an anomalously large correlation for the emission of the second photon.

If $V(R_{ij}) = 0$, similar curves are obtained for $\mathcal{N} > 1$, however when $V(R_{ij}) > \gamma_{EIT}$, Fig. 8.5b shows that interactions cause the two and three atom system to be very strongly bunched at $\tau = 0$, as expected from the simple model in Sect. 8.4. This bunching can be understood from the analytic EIT dark state for the interacting two-atom system in Eq. 5.21, which has a $|ee\rangle$ component in place of the $|rr\rangle$ state expected without interactions. If one of the atoms emits a photon, then $|ee\rangle$ was populated and correspondingly the other atom must emit a photon within a few spontaneous lifetimes. This is a cooperative emission process mediated by the dipole–dipole interactions. The correlation function therefore verifies that the blockade mechanism scatters multiple photons with very high probability. In (c) and (d) the correlations for $\Omega_p = \Gamma_e/2$ and $\Omega_p = \Gamma_e$ are plotted, showing that for a strong probe field the blockade condition is violated and the light becomes anti-bunched at short times, similar to the correlations for the probe-only system in (a).

Figure 8.5b therefore shows that a Rydberg superatom could be used as a correlated photon source. The directionality of the emission is considered below.

8.5.4 Distinguishable Emission

Making a further assumption that the fluorescence emitted by each atom is distinguishable (for example in spatially separated dipole traps as in the experiments in Orsay [21] and Madison [22]) it is possible to also calculate the self- and cross-correlations between atoms i and j using

$$g_{ij}^{(2)}(\tau) = \frac{\langle \hat{\pi}_i^+(t)\hat{\pi}_j^+(t+\tau)\hat{\pi}_j^-(t+\tau)\hat{\pi}_i^-(t)\rangle}{\langle \hat{\pi}_i^+(t)\hat{\pi}_i^-(t)\rangle\langle \hat{\pi}_j^+(t+\tau)\hat{\pi}_j^-(t+\tau)\rangle}, \tag{8.24}$$

which is evaluated in exactly the same way as for $g^{(2)}$ except $\hat{\Pi}^{\pm}$ is replaced by the single-atom dipole operators. The cross-correlation provides an insight into whether the emission from one atom is related to emission of a neighbouring atom.

Figure 8.6 shows the results for the two-atom model calculated for the same parameters as before. In the case of two-level atoms, (a), $g_{21}^{(2)}(\tau) = 1$ for all times as the atoms are independent with no correlations between their emission. This is why the self-correlation shows the same correlation function as for $\mathcal{N} = 1$ in Fig. 8.5a. For the interacting EIT system however, the bunched behaviour is dominated by the cross-correlation, seen from Fig. 8.6b, which is consistent with the interpretation of the bunching as the population of $|ee\rangle$ discussed above. For (b)–(d), the self-correlation remains approximately constant, whilst the cross-correlations change from being strongly bunched to anti-bunched as the probe power is increased.

These results show that it is possible to not only use the strong Rydberg interactions to generate a single-photon output train, but also to obtain highly correlated fluorescence emission from a pair of atoms. In the current assumption of incoherent phase, the direction of the fluorescence will be uncorrelated, however if the phases

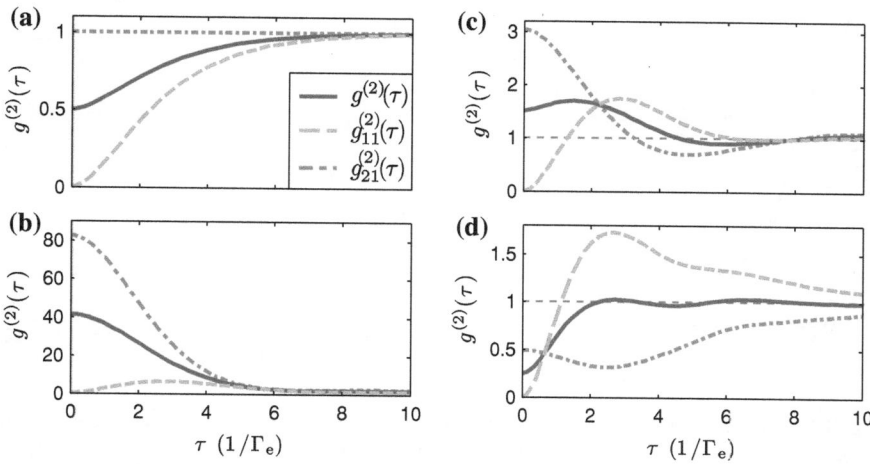

Fig. 8.6 Self- and cross-correlations for $\mathcal{N} = 2$. **a** $\Omega_p = \Gamma_e/5$. Two-level atoms have an independent cross-correlation as the atoms are non-interacting. **b–d** Interacting EIT system with $\Omega_c = \Gamma_e$, $V(R_{ij}) = 2\,\Gamma_e$ and $\Omega_p = \Gamma_e/5$, $\Gamma_e/2$ and Γ_e respectively. This shows the bunching arises from the strong cross-correlation between the atoms, which are correlated by the dipole blockade. For a strong probe, this cross-correlation is suppressed as the system is no longer blockaded

are well defined there exist geometries in which the correlations are insensitive to the atomic position.

8.5.5 Coherent Emission

To check the effects of the the incoherent assumption from above, the correlation function is evaluated using the $e^{\pm i k\hat{\mathbf{r}}\cdot\mathbf{r}_i}$ phase-factors. This also requires a modification of the optical Bloch equations to include the phase of the driving field in the Rabi frequencies as given in Eq. C.4. The correlation function is then calculated for a pair of atoms with the detectors placed orthogonal to the probe wave-vector $\mathbf{k}_p$ as a function of atomic separation in terms of the probe wavelength λ for $\Omega_p = \Gamma_e/2$, $\Omega_c = \Gamma_e$ and $V(R) = 2\,\Gamma_e$.

The results are plotted in Fig. 8.7 which shows the correlations for atoms aligned parallel (a) and perpendicular (b) to the probe beam. For the parallel geometry in (a), the photons are bunched independent of separation R, whilst for the perpendicular configuration in (b) there is a destructive interference for $R = m\lambda + \lambda/4, 3\lambda/4, \ldots$ resulting in anti-bunching. This suggests the geometry of (a) is more robust for observation of photon blockade, and could be used to generate highly correlated photon pairs. For other detector and atom geometries, the correlation function becomes more sensitive to displacement, resulting in more complex correlation functions.

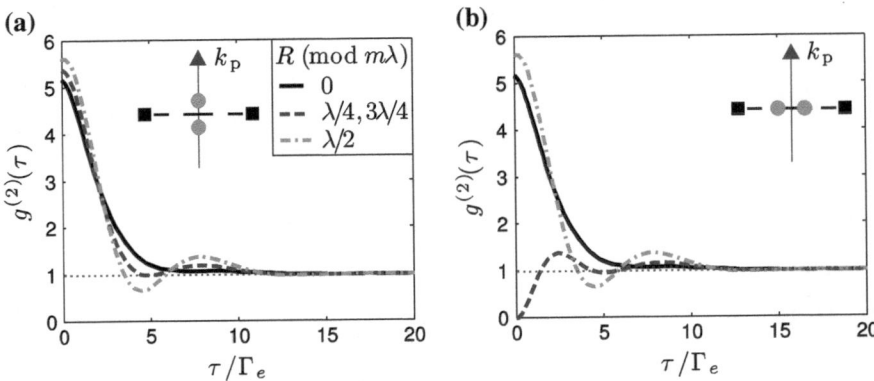

Fig. 8.7 Correlation function for coherent emission for different separations R, where m is an integer. **a** Atoms aligned parallel to the probe beam are bunched for all separations. **b** The perpendicular configuration shows anti-bunching due to destructive interference for $R = m\lambda + \lambda/4, 3\lambda/4$

8.6 Summary

In this chapter the concept of a quantised light-field has been introduced, along with its relevance to generating non-classical light-fields using the blockade effect. Rydberg atom interactions ensure only a single dark state polariton can pass lossless through the blockade region, whilst other photons arriving at the medium will be scattered to achieve photon blockade.

A simple model has been used to predict the correlation function arising from this interaction, which shows that a large optical depth in a single blockade sphere is required to obtain a high fidelity single photon output train. Quantitative calculations of the correlations in the scattered light from a few blockaded atoms verify that the blockade causes the atoms to scatter pairs of photons with very high probability, as seen from the strong bunching in the correlation function at short times. These calculations also highlight the importance of geometry in the system, with photon blockade working better for atoms parallel to the probe to avoid sensitivity to atomic position.

The process considered in this chapter is photons scattered out of the probe beam which is destructive. However, this scattering is conditional on whether another photon is in the medium. This conditional behaviour for the case of one or two photons is a first step towards the development of a two-photon quantum gate, as it shows the blockade mechanism is already sufficient to give a non-linearity at the single photon level. Future work should look for ways to use this effect in the dispersive regime to create a phase-shift on the photons. The first challenge though is to create, and probe, a single blockaded ensemble which has a sufficiently large optical depth. Measurement of the anti-bunching from the photon blockade would enable verification of confinement to $R < R_b$ before moving on to explore the dispersive regime.

References

1. R. Loudon, *The Quantum Theory of Light*, 2nd edn. (OUP, Oxford, 1997)
2. V. Weisskopf, E. Wigner, Berechnung der natürlichen Linienbreite auf Grund der Diracschen Lichttheorie. Z. Phys. **63**(1), 54 (1930)
3. M.O. Scully, M.S. Zubairy, *Quantum Optics* (CUP, Cambridge, 2002)
4. H.C. Carmichael, *Statistical Methods in Quantum Optics 1: Master Equations and Fokker-Planck Equations* (Springer, Berlin, 2002)
5. R. Hanbury Brown, R.Q. Twiss, A test of a new type of stellar interferometer on sirius. Nature **178**, 1046 (1956)
6. H.J. Kimble, M. Dagenais, L. Mandel, Photon antibunching in resonance fluorescence. Phys. Rev. Lett. **39**(11), 691 (1977)
7. F. Diedrich, H. Walther, Nonclassical radiation of a single stored ion. Phys. Rev. Lett. **58**(3), 203 (1987)
8. A. Imamoğlu, H. Schmidt, G. Woods, M. Deutsch, Strongly interacting photons in a nonlinear cavity. Phys. Rev. Lett. **79**(8), 1467 (1997)
9. K.M. Birnbaum, A. Boca, R. Miller, A.D. Boozer, T.E. Northup, H.J. Kimble, Photon blockade in an optical cavity with one trapped atom. Nature **436**, 87 (2005)
10. B. Dayan, A.S. Parkins, T. Aoki, E.P. Ostby, K.J. Vahala, H.J. Kimball, A photon turnstile dynamically regulated by one atom. Science **319**, 1062 (2008)
11. C. Lang, D. Bozyigit, C. Eichler, L. Steffen, J.M. Fink, A.A. Abdumalikov, M. Baur, S. Filipp, M.P. da Silva, A. Blais, A. Wallraff, Observation of resonant photon blockade at microwave frequencies using correlation function measurements. Phys. Rev. Lett. **106**, 243601 (2011)
12. M. Fleischhauer, M.D. Lukin, Dark-state polaritons in electromagnetically induced transparency. Phys. Rev. Lett. **84**(22), 5094 (2000)
13. M. Fleischhauer, A. Imamoğlu, J. Marangos, Electromagnetically induced transparency: optics in coherent media. Rev. Mod. Phys. **77**, 633 (2005)
14. L. Mandel, E. Wolf, *Optical Coherence and Quantum Optics* (CUP, Cambridge, 2008)
15. A. Imamoğlu, Y. Yamamoto, Turnstile device for heralded single photons: Coulomb blockade of electron and hole tunneling in quantum confined p-i-n heterojunctions. Phys. Rev. Lett. **72**(2), 210 (1994)
16. A.J. Shields, Semiconductor quantum light sources. Nature Photon. **1**, 215 (2007)
17. M. Born, E. Wolf, *Principles of Optics* (CUP, Cambridge, 1999)
18. G.S. Agarwal, Quantum statistical theories of spontaneous emission and their relation to other approaches. Springer Tracts Mod. Phys. **70**, 1 (1974)
19. K. Mølmer, Correlation functions and the quantum regression theorem. http://owww.phys.au.dk/quantop/kvanteoptik/qrtnote.pdf. Accessed 21 Nov 2010
20. K. Mølmer, Y. Castin, Monte Carlo wavefunctions in quantum optics. Quantum Semiclass Opt. **8**(1), 49 (1996)
21. A. Gaëtan, Y. Miroshnychenko, T. Wilk, A. Chotia, M. Viteau, D. Comparat, P. Pillet, A. Browaeys, P. Grangier, Observation of collective excitation of two individual atoms in the Rydberg blockade regime. Nat. Phys. **5**, 115 (2009)
22. E. Urban, T.A. Johnson, T. Henage, L. Isenhower, D.D. Yavuz, T.G. Walker, M. Saffman, Observation of Rydberg blockade between two atoms. Nat. Phys. **5**, 110 (2009)

Chapter 9
Progress Towards a Single Blockade Sphere

9.1 Design Constraints

In the previous chapter photon blockade was considered to illustrate the ability to realise non-classical states of light using the strong dipole–dipole interactions of the Rydberg states. From the results of the simple model in Sect. 8.4, two key requirements were determined to enable creation of a highly correlated single-photon output;

1. A single blockade region in the transverse mode of the probe laser with waist $w_0 < R_b/2$ to prevent formation of more than a single dark state polariton,
2. An optical depth OD $\gtrsim 4$ to maximise the probability of scattering photons out of the probe beam.

An implicit assumption relevant to both of these requirements is that the atomic cloud has a comparable size to the probe waist to ensure that light can be scattered from the edges of the probe beam whilst maintaining the blockade condition $R < R_b$ across the sample.

Typically $R_b \sim 5\,\mu\text{m}$ (although this can be increased by choice of n), requiring a probe beam focus around $1\,\mu\text{m}$ to meet the first constraint. This waist corresponds to $>98\,\%$ probe beam intensity contained within a diameter R_b. For a perfect lens with numerical aperture NA, the smallest possible focus is given by the Airy radius $r = 1.22\lambda/2\text{NA}$. Comparing this to the $1\,\mu\text{m}$ waist for the 780 nm probe laser, a diffraction limited lens with NA ~ 0.5 is needed.

In the apparatus detailed in Part II there is insufficient optical access to obtain a diffraction limited probe waist. A new experiment has therefore been designed and built, which is described in the sections below.

J. D. Pritchard, *Cooperative Optical Non-Linearity in a Blockaded Rydberg Ensemble*,
Springer Theses, DOI: 10.1007/978-3-642-29712-0_9,
© Springer-Verlag Berlin Heidelberg 2012

9.2 Trapping Atoms in a Single Blockade Volume

For the second requirement of large optical depth in the blockade volume, it is insufficient to simply probe atoms in the MOT as the density is too low, as discussed in Sect. 8.4. Instead, atoms must be loaded into an optical dipole trap which typically gives a density of 10^{12} cm^{-3} [1, 2]. Using a dipole trap coaxial with the probe beam has the advantage of providing tight transverse confinement of the atoms to a dimension smaller than the blockade radius. It can also be focussed using the same optics as the probe laser.

9.2.1 Dipole Force

The dipole trapping force arises from the AC Stark shift of an atom driven by a far detuned laser field, which creates a potential $U \propto I/\Delta$ where I is the laser intensity [3]. This creates a conservative force $\mathbf{F} = -\nabla U$ that is proportional to the gradient of the potential, where the sign depends on the detuning of the laser. The atomic dipole also has a component out of phase with the driving field, and this causes atoms to scatter photons from the trapping laser at a rate $\propto I/\Delta^2$. This scattering heats atoms out of the trap, so a large Δ is desirable, requiring increased laser intensity to maintain trap depth.

For a red detuned laser ($\Delta < 0$), the atom is trapped at the point of highest intensity, and atoms can be confined using a focused Gaussian beam with waist w_0. The resulting trap potential is given by [1]

$$U(r, z) = \frac{U_0 \exp\{-2r^2/w(z)^2\}}{1 + (z/z_R)^2}, \tag{9.1}$$

where $z_R = \pi w_0^2/\lambda$ is the Rayleigh range, $w(z) = w_0\sqrt{1 + (z/z_R)^2}$ and U_0 is the trap depth. Approximating the trap to a harmonic potential well, the density distribution in the trap is described by a 3D Gaussian. The radii of this distribution are $\sigma_r = \sqrt{k_B T w_0^2/4U_0}$ and $\sigma_z = \sqrt{k_B T z_R^2/2U_0}$ in the transverse and longitudinal directions respectively [1], where T is the temperature. Typically, the atoms thermalise in the trap at a temperature of $k_B T \sim U_0/10$ [2, 4]. Using this empirical factor, the spatial extent of the cloud in the trap is around $\sigma_r \sim 0.16w_0$, $\sigma_z \sim 0.2z_R$, resulting in a very tight transverse confinement but an elongated sample along the beam axis.

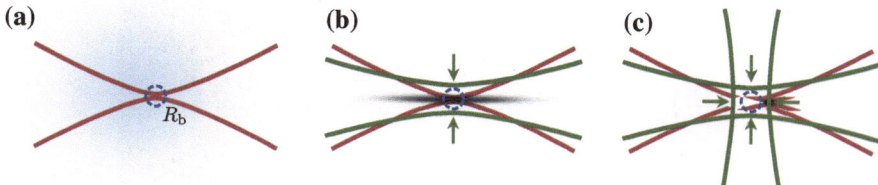

Fig. 9.1 Dipole trap geometry. **a** Focusing the probe down to $1\,\mu$m in the MOT gives almost no absorption, as the density is too low; **b** a $5\,\mu$m dipole trap along the probe axis provides tight transverse confinement $< R_b$, however the longitudinal cloud size equivalent to around $6R_b$; **c** adding a $6\,\mu$m transverse beam to create a crossed dipole trap allows complete 3D confinement within a blockade sphere

9.2.2 Dipole Trap Setup

For the new experiment, a wavelength of 915 nm is used for the dipole trap laser, which is derived from a home-built TA system giving 1 W output. For the transverse confinement, a waist of $w_0 = 5\,\mu$m is used to obtain a cloud size comparable to the probe waist to give good mode matching whilst confining atoms within a blockade radius. This waist corresponds to a Rayleigh range of $z_R = 86\,\mu$m, resulting in a longitudinal cloud length of $\sim 35\,\mu$m, much larger than the blockade radius as shown schematically in Fig. 9.1b. It is therefore necessary to use a crossed dipole trap configuration, where a second laser beam is used to provide longitudinal confinement along the probe axis.

The high numerical aperture required to obtain a tight probe focus places a number of constraints on the optical access perpendicular to the probe beam. The horizontal MOT beams cross at an angle of 20°, which leaves an effective NA ~ 0.12 for the cross-trap. Using Zemax optical modelling software, a multi-element lens config-uration has been designed to give a $6\,\mu$m Gaussian waist, requiring a high optical quality viewport. Using this second laser, the atoms can be confined to a longitudinal radius of $1\,\mu$m, enabling confinement within a single blockade volume, as illustrated in Fig. 9.1c. This second beam must be orthogonally polarised to the first to prevent creating unwanted effects due to interference.

There are a number of advantages to using the cross-trap geometry. The increased volume of the larger dipole trap can be used to load a greater number of atoms into the dipole trap. This can be used as a reservoir of atoms for enhanced loading of the tighter trap, and may enable techniques such as evaporative cooling [5] with a dimple trap [6, 7] to be used to increase the density of atoms in the blockade region. It also allows the relative cloud size, and hence aspect ratio of the cloud, to be varied by changing the relative powers in the trapping lasers. This could be useful for exploring the angular dependence of the emission from a single blockade volume [8].

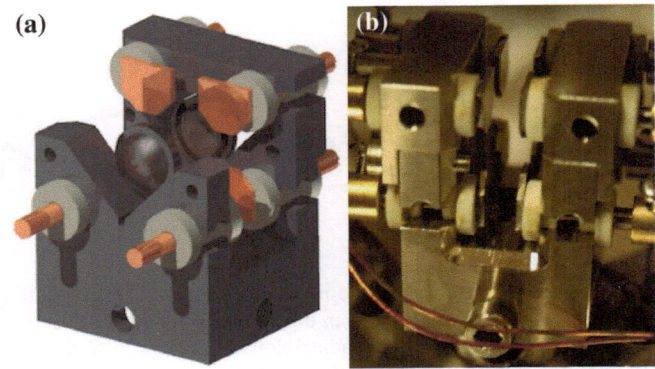

Fig. 9.2 V-block designed for mounting the aspheric lenses in vacuum. The lenses are surrounded by electrodes which enable cancellation of stray fields in all axes, which are insulated from the V-block using ceramic spacers (*white*). **a** CAD model showing lenses aligned in V-block; **b** actual V-block mounted in experiment. Holes in the block prevent virtual leaks from trapped air pockets

9.3 Experiment Setup

9.3.1 Diffraction Limited Optics

There are a number of options for achieving a diffraction limited imaging system with $NA = 0.5$, such as commercial microscope objectives or custom multi-element lens configurations [9, 10]. Typically these components are not compatible with ultra-high vacuum, and must focus light tightly through a vacuum window which can introduce aberrations into the system. The alternative is to place a diffraction limited aspheric lens inside the vacuum system, greatly simplifying the optical design. This approach has been successfully demonstrated in a number of groups for creating microscopic dipole traps in which only a single-atom can be loaded [11, 12], enabling measurement of 10 % absorption from a single atom [13].

For the new experiment, a pair of aspheric lenses manufactured by Lightpath Technologies, Inc. (catalogue number 350240) are used to focus and recollect the probe beam. These lenses have been chosen following their extensive characterisation for applications in single atom trapping [14]. The aspherics have $NA = 0.5$ and are designed to be diffraction limited at 780 nm for a collimated input beam if a 0.25 mm glass window is placed in front of the lens. However, if this window is absent, diffraction limited performance can be restored by using a weakly convergent input beam.

The relative alignment of the lenses in the vacuum chamber is crucial. It not only affects the collection efficiency of the probe laser after the focus, in future experiments it may be required to use counter-propagating probe lasers with overlapped foci. A symmetric alignment is therefore needed. The lenses are glued onto a custom V-block machined from 316LN stainless steel, shown in Fig. 9.2. The V-block defines the optical axis for the two lenses, however the lenses can still be tilted relative to this axis. These degrees of freedom were set by careful alignment of the lenses using a commercial Shack-Hartmann interferometer to measure the wavefront curvature and distortions, performed by A. Gauguet. The first lens is aligned to be perpendicular relative to a well-defined input beam. The lens is then clamped in place and the glued using UHV compatible Epotek H77. To cure the glue, the V-block must be heated to 135 °C for 4 h in an oven. Once cooled, the V-block is placed back in the input beam and the second lens adjusted to match the wave-front curvature of the input beam using a three-axis translation stage. The second lens is then clamped and glued. After curing, the peak and rms wave-front errors added to the input beam by propagation through the aspheric lens pair are measured to be $\lambda/5$ and $\lambda/20$ respectively, showing this setup gives very good relative alignment of the two lenses. Reversing the lenses results in the same magnitude of wave-front errors, demonstrating the symmetric and reversible alignment achieved using this method. As a further test, the assembled V-block was additionally heated to 140 °C for 6 h without any clamps to ensure the alignment would survive the vacuum bakeout in the chamber. Repeated tests with the Shack-Hartmann reveal no change in the alignment from the repeated baking. Finally, a piezo-electric translation stage was used to knife-edge the probe focus, obtaining a waist of $1.2 \pm 0.1\,\mu m$ using a beam with a 6 mrad convergence angle and a $1/e^2$ waist of 2 mm at the first lens face. Tighter foci are possible using larger input beams, however this results in loss due the lens aperturing the beam.

9.3.2 Electric Field Control

As discussed in Sect. 2.4, Rydberg states have extreme electric field sensitivity. It is therefore necessary to be able to cancel stray-fields around the atoms or to apply a well defined electric field, e.g. to change to $1/R^3$ interactions using a Förster resonance. To provide electric field control, four polished steel electrodes are mounted around each lens, as shown in Fig. 9.2, which enables electric fields to be applied or cancelled on all three axes. These are insulated from the grounded V-block using ceramic spacers. The electrodes and spacers are glued onto the V-block using Epotek H77.

One of the biggest sources of stray-field in this setup could come from patch-potentials building up on the surface of the lenses over time, for example due to deposition of rubidium atoms and ions [15]. The lenses have therefore been coated with a conductive indium tin oxide (ITO) layer in addition to an anti-reflection (AR) coating for 780 and 480 nm, preventing charge build up on the lens. The disadvantage of this layer is that its refractive index cannot be well matched to provide a low reflection AR coating, resulting in a transmission of 90 % through each lens at 780 nm. It also limits the bakeout temperature to 150 °C to prevent degradation of the ITO from reaction with oxygen in the air. The electrical contact is made using a piece of solder (see Appendix B) clamped onto the top of the lens by the mounting bar for the electrodes above each lens.

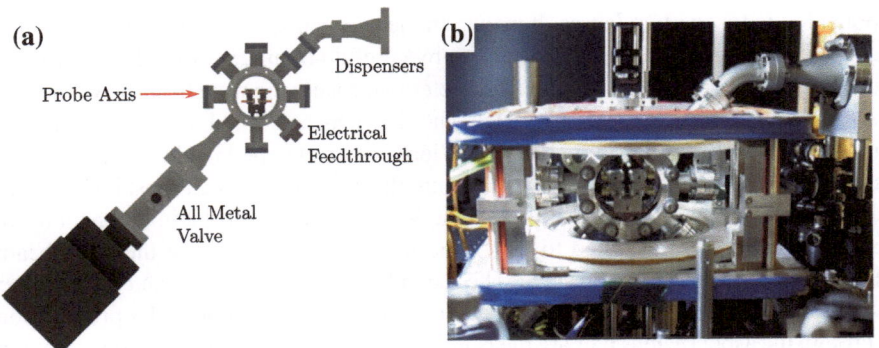

Fig. 9.3 New experiment vacuum chamber. **a** The V-block is mounted in a pancake shape vacuum chamber, with the lens axis aligned horizontally; **b** assembled chamber including magnetic coils. The cage visible above the chamber delivers the vertical MOT beam to the chamber

9.3.3 Vacuum Chamber

The V-block is mounted at the centre of a pancake-shaped chamber, as shown in Fig. 9.3a, with the probe axis aligned to be in the horizontal plane. The two large viewports at the side of the chamber are constructed using the method detailed in Appendix B to provide high optical quality viewport windows close to the edges of the V-block. These windows minimise the aberrations induced on the cross-trap, and provide good optical access. There are eight DN16CF flanges around the edge of the chamber, with the vertical and horizontal pairs used for the vertical MOT beam and probe axis respectively. A pair of Alvasource dispensers are mounted diagonally above the lens axis. These are contained in a conical reducer with a 45° bend to avoid direct line of sight from the dispensers onto the lenses. This prevents the lenses being coated with rubidium, which will cause them to become opaque. The two lower 45° flanges are used for an electrical feedthrough for the eight electrodes and the Gamma Vacuum Titan 20S ion pump, which has a 20 l/s pumping speed. This is connected via a T-piece with an all-metal valve to seal the chamber off after pumping down. The chamber was baked at 150 °C to enable pumping down to a vacuum of 10^{-11} torr, measured using an ion gauge.

9.3.4 Beam Alignment

Alignment of the probe beam through the chamber requires very high tolerances on the matching of the input and output convergence and beam waist to obtain a reversible optical path with a focus as the centre of the chamber. As the probe beam fills the clear aperture of the lens, it must also be well centred on the optical axis of the two lenses to prevent clipping of the beam edges, which will introduce aberrations and limit the transmission through the system.

The probe beam is set up using a fixed cage-mount system to expand the light from a bare single mode polarisation maintaining (SPM) fibre, as shown in Fig. 9.4a. The beam is weakly focused using an achromat to obtain a convergence angle of 6.48 ± 0.07 mrad, measured by knife-edging along the beam over a 1 m path length and fitting to a Gaussian beam profile. The probe is aligned into the chamber using caps on the viewports to provide mechanical alignment onto the optical axis of the lenses. The divergence of the output beam is then measured by profiling the beam with a minimum of 5 knife-edge measurements at 10–20 cm separations. The distance between the input cage and the first aspheric lens in the chamber is then adjusted until the output beam matches the input, with a divergence angle of 6.49 ± 0.1 mrad. The output beam is then coupled back into a second SPM fibre with an identical cage setup, adjusting the alignment to give a reproducible and reversible coupling between both fibres. This output beam is connected to the SPAD to record probe transmission. Using the input beam convergence angle and $1/e^2$ beam waist of 2.17 ± 0.07 mm on the first lens, Zemax lens modelling software was used to find the effective focal length of the lens pair as 5.63 ± 0.03 mm.

The desired waist for the longitudinal dipole trap is 5 μm. Using the effective focal length from the probe measurement, Zemax is used to find the input beam waist and convergence angle required to achieve this spot size at the position of the probe focus. The dipole trap light is then set to a convergence angle of 1.93 ± 0.04 mrad using an adjustable focal length collimator, and aligned in the same manner as the probe beam to match the input and output convergence. The final alignment corresponds to an input beam waist of 0.45 ± 0.02 mm, giving a calculated focus of 5.0 ± 0.2 μm inside the chamber. The probe and dipole trap light are combined using a dichroic mirror before the chamber, which allows both beams to be coupled into the output fibre to provide transverse alignment of the beams. An interference filter is then added to prevent dipole trap light reaching the SPAD. Similarly, the 480 nm light is aligned into the chamber to counter-propagate with the probe beam. This is set to give a spot size of 18 μm inside the chamber, ensuring a uniform illumination of the blockade region to prevent Ω_c changing across the sample.

9.3.5 MOT Alignment

Due to the high NA of the aspheric lenses, it is not possible to use the standard MOT geometry with three pairs of orthogonal beams. Instead, the horizontal MOT beams cross at an angle of $20°$. The vertical MOT beam is coupled into a polarisation maintaining fibre and expanded to a $1/e^2$ waist of 2.4 mm using a cage mount on the top of the chamber, as seen in Fig. 9.3b. The horizontal beams have a waist of 1.4 mm to ensure they do not clip the electrodes mounted around lenses. The MOT coils and three axis bias coils are mounted on a frame around the chamber, which provides an additional degree of freedom, combined with the beam alignment, with

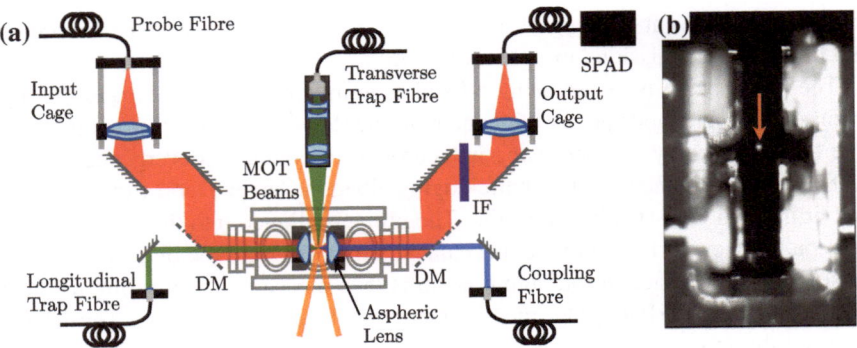

Fig. 9.4 New experiment setup. **a** Schematic of beam alignment around chamber. Light for dipole trapping and Rydberg excitation is overlapped with the probe beam using dichroic mirrors (DM), and filtered out of the collection optics using interference filters (IF); **b** image of MOT at the centre of the V-block. The electrodes and ceramic spacers are clearly visible around the atoms

which the MOT can be positioned at the centre of the lens axis. This is crucial, as the probe beam is aligned to be in the centre of the two lenses, and the dipole traps only collect atoms from a region equivalent to the Rayleigh range. Thus, even for the larger transverse dipole trap, the MOT must be aligned onto the centre to better than around 100 μm.

The MOT has the best loading at a field gradient of 15 G/cm for cooling light detuned by $\Delta = -2.3\,\Gamma_e$ at an intensity of 0.35 mW/cm^2 in each beam. For a load of 1 s, around 500,000 atoms are collected in the MOT with a cloud size ∼0.1 mm. An image of the MOT at the centre of the V-block can be seen in Fig. 9.4b. The position of the MOT is very sensitive to beam balance and magnetic field cancellation around the chamber. Further optimisation of the alignment and position is required to improve the stability and reproducibility of the atoms in the lens axis.

9.4 Summary and Outlook

In this chapter the construction of a new experiment for the observation of non-classical light from interaction with a single blockade volume is described. The initial steps towards localising atoms within a single blockade sphere have been made, with the probe and dipole traps aligned into the chamber and cold atoms being collected in the MOT. One of the major difficulties in building the chamber was to develop a method to mount the aspheric lenses in vacuum. Gluing the lenses onto the V-block using the Shack-Hartmann, as described above, provides a reliable technique for obtaining the required robust, symmetric alignment of the lenses.

Given the opportunity to build this apparatus again, alternative lenses would be chosen to enable use of a collimated probe beam. This reduces the tolerances required for the alignment of the input beams, and is better suited for experiments seeking to measure the phase-shift of the probe by building an interferometer around the setup. Currently, the weak divergence of the probe beam makes mode-matching with a reference beam challenging, potentially requiring a second aspheric lens pair to be aligned in the reference arm [16]. The ITO coating would also be neglected in a future setup as there are already dark-spots appearing on the lenses from where rubidium has reacted with the coating. Despite these suggested improvements, the current apparatus meet the requirements for probing a single blockade region and will provide an excellent test-bed for looking for single-photon non-linearities. The next steps will be to begin optimising the loading of the dipole trap to get a large optical depth in the blockade volume.

Once obtained, an optically thick, isolated ensemble of atoms confined within a blockade radius opens the possibility of studying a rich variety of non-classical states of light in addition to photon blockade, such as generating single-photons using four-wave mixing [17, 18] or photon-subtracted states [19]. This setup should provide a flexible and versatile apparatus with which to characterise the blockade mechanism, as there is the possibility of using different trap geometries and also the ability to tune the interactions with electric fields. Dipole blockade has currently only been demonstrated in macroscopic ensembles or for isolated atom pairs, however probing a single ensemble allows direct testing of the collective blockade state. This is important not only for experiments to generate and manipulate light at the single photon level, but also for proposals to utilise the collective nature of blockade to build atomic quantum gates [20, 21].

References

1. R. Grimm, M. Weidemuller, Y.B. Ovchinnikov, Optical dipole trap for neutral atoms. Adv. At. Mol. Opt. Phys. **42**(95), 170 (2000)
2. C.S. Adams, E. Riis, Laser Cooling and trapping of neutral atoms. Prog. Quant. Electron. **21**(1), 1 (1997)
3. C.J. Foot, *Atomic Physics* (OUP, Oxford, 2005)
4. J.D. Miller, R.A. Cline, D.J. Heinzen, Far-off-resonance optical trapping of atoms. Phys. Rev. A **47**(6), R4567 (1993)
5. C.S. Adams, H.J. Lee, N. Davidson, M. Kasevich, S. Chu, Evaporative cooling in a crossed dipole trap. Phys. Rev. Lett. **74**(18), 3577 (1995)
6. T. Weber, J. Herbig, M. Mark, H.-C. Nägerl, R. Grimm, Bose-Einstein condensation of cesium. Science **299**, 232 (2003)
7. Z.Y. Ma, C.J. Foot, S.L. Cornish, Optimized evaporative cooling using a dimple potential: an efficient route to Bose-Einstein condensation. J. Phys. B **37**, 3187 (2004)
8. L.H. Pedersen, K. Mølmer, Few qubit atom-light interfaces with collective encoding. Phys. Rev. A **79**(1), 012320 (2009)
9. N. Schlosser, Étude et réalisation de micri-piéges dipoliares optiques pour atomes neutres. Ph.D. Thesis, Université Paris XI, Orsay, 2002
10. W. Alt, An objective lens for efficient fluorescence detection of single atoms. Optik **113**(3), 142 (2002)
11. N. Schlosser, G. Reymond, P. Grangier, Collisional blockade in microscopic optical dipole traps. Phys. Rev. Lett. **89**(2), 023005 (2002)

12. M. Weber, J. Volz, K. Saucke, C. Kurtsiefer, H. Weinfurter, Analysis of a single-atom dipole trap. Phys. Rev. A **73**(4), 043406 (2006)
13. M.K. Tey, Z. Chen, S.A. Aljunid, B. Chng, F. Huber, G. Maslennikov, C. Kurtsiefer, Strong interaction between light and a single trapped atom without a cavity. Nat. Phys. **4**, 924 (2008)
14. Y.R.P. Sortais, H. Marion, C. Tuchendler, A.M. Lance, M. Lamare, P. Fournet, C. Armellin, R. Mercier, G. Messin, A. Browaeys, P. Grangier, Diffraction-limited optics for single-atom manipulation. Phys. Rev. A **75**, 013406 (2007)
15. A. Tauschinsky, R.M.T. Thijssen, S. Whitlock, H.B. van Linden van den Heuvell, R.J.C. Spreeuw, Spatially resolved excitation of Rydberg atoms and surface effects on an atom chip. Phys. Rev. A **81**, 063411 (2010)
16. S.A. Aljunid, M.K. Tey, B. Chng, T. Liew, G. Maslennikov, V. Scarani, C. Kurtsiefer, Phase shift of a weak coherent beam induced by a single atom. Phys. Rev. Lett. **103**(15), 153601 (2009)
17. M. Saffman, T.G. Walker, Creating single-atom and single-photon sources from entangled atomic ensembles. Phys. Rev. A **66**, 065403 (2002)
18. E. Brekke, J.O. Day, T.G. Walker, Four-wave mixing in ultracold atoms using intermediate Rydberg states. Phys. Rev. A **78**(6), 063830 (2008)
19. J. Honer, R. Löw, H. Weimer, T. Pfau, H.P. Büchler, Artificial atoms can do more than atoms: deterministic single photon subtraction from arbitrary light fields. Phys. Rev. Lett. **107**, 093601 (2011)
20. M.D. Lukin, M. Fleischhauer, R. Cote, L.M. Duan, D. Jaksch, J.I. Cirac, P. Zoller, Dipole blockade and quantum information processing in mesoscopic atomic ensembles. Phys. Rev. Lett. **87**(3), 037901 (2001)
21. M. Müller, I. Lesanovsky, H. Weimer, H.P. Büchler, P. Zoller, Mesoscopic Rydberg gate based on electromagnetically induced transparency. Phys. Rev. Lett. **102**(17), 170502 (2009)

Part IV
Conclusions and Outlook

Chapter 10
Conclusion

In this thesis, Rydberg EIT has been used to combine the strong dipole–dipole interactions of the Rydberg states with the resonant *Dark state* to realise a novel cooperative optical non-linearity. An interacting $\mathcal{N}$-atom model was developed to show the effect of the dipole blockade is to prevent more than a single *Dark state* in each blockade region. The remaining atoms scatter photons from the probe laser, suppressing the resonant transmission.

Experiments have been performed on a cold atomic ensemble to look for evidence of interaction effects using EIT for Rydberg atoms with $n = 19 - 60$. For states with $n \lesssim 26$, interactions are manifested as a density-dependent loss, consistent with superradiance, that dominates over the suppression mechanism due to the geometric enhancement from the atom cloud diameter compared to the emission wavelength.

At $n \sim 60$, the cooperative optical non-linearity has been observed and characterised for both attractive and repulsive dipole–dipole interactions. Results for the repulsive interactions conclusively rule out alternative mechanisms for the suppression, and excellent quantitative agreement is obtained at low density to the three-atom model, placing an upper limit of 110 kHz on the relative dephasing rate between neighbouring blockade spheres. Attractive interactions result in a non-linearity dependent upon the direction of the frequency scan, characterised by second-(third-)order non-linear susceptibilities for scanning across the two-photon resonance with an initially positive (negative) probe frequency. For both directions, the magnitude of the non-linear susceptibility is significant when compared to other non-linear media, and a quadratic density dependence consistent with cooperativity is observed.

One of the limitations of the data presented in Sect. 7.3 for attractive interactions is the lack of information about the ion fraction to complement the transmission spectra. Using an MCP and electrodes in the vacuum chamber, the attractive regime could be studied further to give insight into the underlying mechanism for the second or third order non-linearities. This would also allow the universal scaling predicted by Ates et al. [1] to be tested.

The blockade mechanism allows a single blockaded ensemble to transmit a single photon through formation of a single-photon *Dark state* polariton whilst scattering

J. D. Pritchard, *Cooperative Optical Non-Linearity in a Blockaded Rydberg Ensemble*,   147
Springer Theses, DOI: 10.1007/978-3-642-29712-0_10,
© Springer-Verlag Berlin Heidelberg 2012

additional photons out of the probe mode. The concept of photon blockade was introduced in Chap. 8 to show the single-photon character of the observed cooperative optical non-linearity, and a model developed to predict the correlation function of the probe after a single blockade region. This can be used to generate a highly correlated train of photons separated in time by $\tau_b \sim 100$ ns with several hundred atoms confined within a single blockade sphere. Equivalently, the blockade can be used to create a highly correlated photon pair source by collecting light scattered from the side of the blockade region.

Progress towards obtaining a single blockade region has been presented, describing construction of a new apparatus in which the requirements of the photon blockade can be realised. This new setup allows studies of a wide range of novel and interesting physics relevant to realising optical non-linearities on the single photon level. Two key areas of future study are;

- **Collective single photon emission** Blockade allows excitation of a collective wavefunction with the Rydberg excitation shared across the ensemble, as discussed in Sect. 5.3. If this excitation is mapped onto the intermediate excited state following a π-pulse, the result is collective emission of a highly collimated single photon at a superradiant rate [2–4]. This process gives enhanced coupling between a single photon and an atomic ensemble, which could allow high-fidelity transport of quantum information between spatially separated ensembles [4]. The cross-trap geometry in the future will enable studies of emission for different aspect ratios of the atomic cloud.
- **Single-photon phase-shift** All of the experiments presented in this thesis are performed through measurements of the transmission. Building an interferometer with the blockade region in one arm will allow measurement of the single-photon phase-shift in the dispersive regime. This can be done using homodyne detection [5] in which the amplitude and phase of the single photon are amplified by a strong reference beam. This setup also allows tomographic reconstruction of the light-field [6], allowing better characterisation of the non-classical light.

Observation and characterisation of the blockade in a single ensemble is important not only for developing photonic devices, but also for other approaches to quantum information processing. In the long term, it may be possible to combine these approaches to develop a high-fidelity photonic phase gate.

References

1. C. Ates, S. Sevinçli, T. Pohl, Electromagnetically induced transparency in strongly interacting Rydberg gases. Phys. Rev. A **83**(4), 041802 (2011)
2. M. Saffman, T.G. Walker, Creating single-atom and single-photon sources from entangled atomic ensembles. Phys. Rev. A **66**, 065403 (2002)
3. I.E. Mazets, G. Kurizki, Multiatom cooperative emission following single-photon absorption: Dicke-state dynamics. J. Phys. B **40**(6), F105 (2007)

4. L.H. Pedersen, K. Mølmer, Few qubit atom-light interfaces with collective encoding. Phys. Rev. A **79**(1), 012320 (2009)
5. M.O. Scully, M.S. Zubairy, *Quantum Optics* (CUP, Cambridge, 2002)
6. G. Breitenbach, S. Schiller, Homodyne tomography of classical and non-classical light. J. Mod. Opt. **44**, 2207 (1997)

Appendix A
Useful Circuits

A.1 Fast Photodiode

The photodiode is designed using a Hamamatsu S5972 500 MHz photodiode combined with a Texas Instruments LT1222 op-amp, which has a 1 GHz gain bandwidth product. This circuit gives a gain of $2:4 \times 10^3$ V/W with a 15 MHz bandwidth, which is used for the modulation transfer lock of the cooling laser, as discussed in Sect. 6.1.1. Best performance is obtained by placing a grounded guard-rail around the non-inverting input (dashed line) and minimising the distance between the chip and the photodiode (Fig. A.1).

A.2 SPAD Protection Circuit

Protection circuit for the Perkin-Elmer SPCM-AQR photon counters. This ensures the detectors are gated for a low TTL, preventing damage from overexposure of the SPAD. The counter can be enabled using an active high Gate TTL. This transistor is chosen based on the more detailed protection circuit in [1] (Fig. A.2).

J. D. Pritchard, *Cooperative Optical Non-Linearity in a Blockaded Rydberg Ensemble*,   151
Springer Theses, DOI: 10.1007/978-3-642-29712-0,
© Springer-Verlag Berlin Heidelberg 2012

Fig. A.1 Fifteen MHz band-width fast photodiode. *Dashed line* denotes grounded guard loop around photodiode anode and non-inverting input on the op-amp to prevent the parasitic oscillating

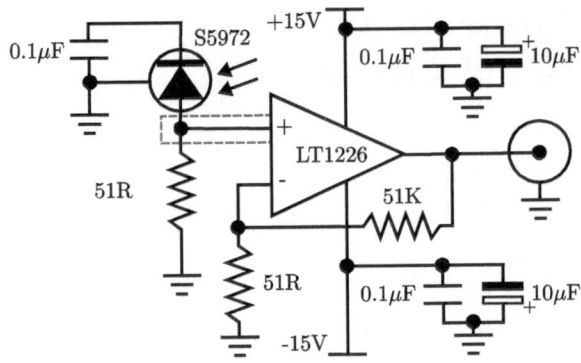

Fig. A.2 SPAD protection circuit to gate by default

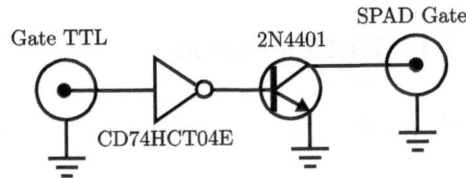

Appendix B
Home-Made Viewport Construction

For the new experiment, two large high-quality vacuum windows were used to provide good optical access from the side of the lens axis. The viewport construction method builds on the design detailed in [2], using a soft solder seal between the glass and metal to prevent stress, and hence birefringence, on the glass window which is typical for standard conflat viewport windows. The difference here is that the windows are now sealed directly onto the vacuum chamber, which reduces the physical size of the viewport and enables the width of the chamber to be kept small.

The windows are high quality BK7 glass with dimensions $\varnothing 70 \times 9$ mm, which are AR coated for 780 and 480 nm with a 5 mm mask around the edge of the lens. This ensures the seal is made directly onto the glass, as placing the solder on the AR coating can reduce the reliability of the vacuum seal. The chamber is designed with a flat rim 2 mm thick, with an outer diameter of 70 mm to match the glass. Around this rim, eight lugs are welded onto the chamber which have an M4 thread to hold the window in place. A cross-section of the viewport is shown in Fig. B.1a, showing the window is clamped onto the chamber using an external flange with a pair of solder

The solder seals are constructed using Indium Corporation WIREOT-51831 $\varnothing = 0.030$" alloy wire (97.5 % Pb, 1.5 % Ag and 1 % Sn). Wire is wrapped round a metal former and soldered into a ring the same diameter as the chamber rim, shown in Fig. B.1b. The ring is then pressed flat to a thickness of 0.3 mm, leaving excess around the solder join Fig. B.1c. This must be inspected to ensure the solder join is not visible, as if the join can be seen the seal will be compromised. The excess solder is then removed using a scalpel. All tools must be clean to prevent getting grease inside the vacuum chamber.

Finally, the window is assembled as shown in Fig. B.1d. The outer flange is secured to the chamber using M4 bolts, with two conical disk springs placed baseto- base as washers on each bolt, which are tightened to a torque of 2 N m. The chamber is then baked at 150 °C (limited by the lens ITO coating) and pumped

J. D. Pritchard, *Cooperative Optical Non-Linearity in a Blockaded Rydberg Ensemble,*   153
Springer Theses, DOI: 10.1007/978-3-642-29712-0,
© Springer-Verlag Berlin Heidelberg 2012

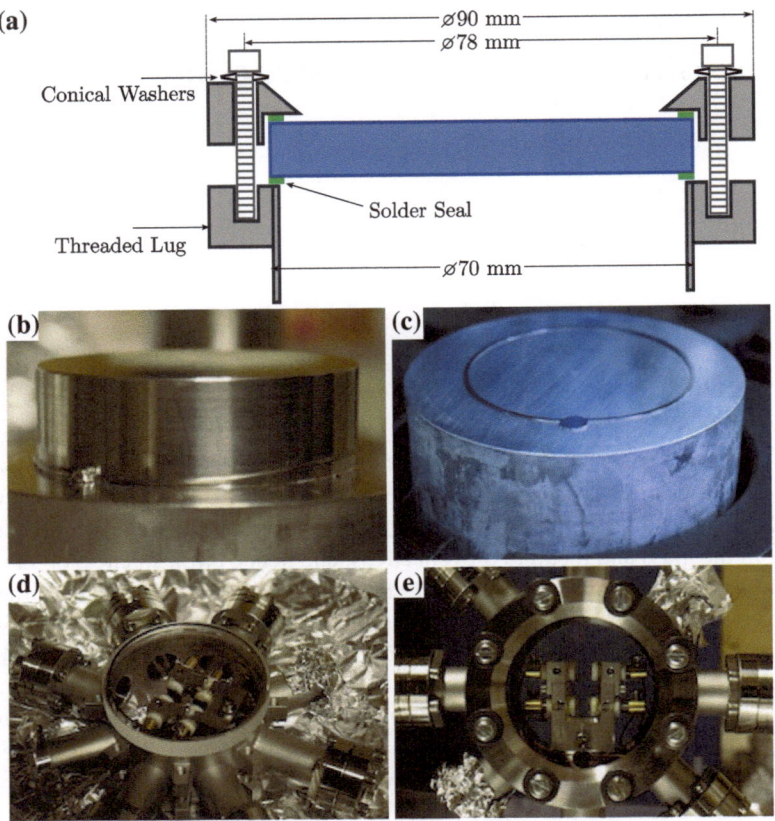

Fig. B.1 Home-made viewport assembly. **a** Cross-section showing outer flange clamping the glass onto the rim of the chamber using two soft solder seals. **b** Solder wire is soldered into a ring using a metal former. **c** The ring is then pressed to a thickness of 0.3 mm to make the seal. **d** Solder seal on glass window. **e** Finished window with outer flange tightened to a torque of 2 N m

down to a pressure of 10^{-11} torr. During baking, the solder rings become soft, however the conical disk springs maintain a force on the flange to prevent leaks. Once cooled, the bolts should be retightened to the original torque. However, the windows on the new chamber remain leak-tight even if the outer flanges are removed, demonstrating the robust seal formed during baking. This design is poorly suited to chambers requiring repeated access to the chamber, as the windows cannot be removed without risking damage to the rim of the chamber, but is a very simple technique for obtaining high quality viewport windows.

Appendix C
Quantised Atom-Light Interactions

Consider a two-level atom at position r_A interacting with modes of the quantised electromagnetic field, as introduced in chapter 8. The Hamiltonian for the coupled system is given by $\hat{\mathscr{H}} = \hat{\mathscr{H}}_A + \hat{\mathscr{H}}_E + \hat{\mathscr{H}}_I$, where each of these terms represents the energy of the bare atom, the energy of the quantised field and the interaction between the atom and the field respectively. Applying the rotating wave approximation, the Hamiltonians for this system are given by [3]

$$\hat{\mathscr{H}}_A = \hbar\omega_0\hat{\pi}^+(t)\hat{\pi}^-(t), \tag{C.1a}$$

$$\hat{\mathscr{H}}_E = \sum_{\mathbf{k}} \hbar\omega_{\mathbf{k}}\hat{a}_{\mathbf{k}}^{\dagger}(t)\hat{a}(t), \tag{C.1b}$$

$$\hat{\mathscr{H}}_I = i\sum_{\mathbf{k}} \hbar g_{\mathbf{k}}\left\{\hat{\pi}^+(t)\hat{a}_{\mathbf{k}}(t)e^{i\mathbf{k}\cdot\mathbf{r}_A} - \hat{a}_{\mathbf{k}}^{\dagger}(t)\hat{\pi}^-(t)e^{-i\mathbf{k}\cdot\mathbf{r}_A}\right\}, \tag{C.1c}$$

where $\hat{\pi}^{\pm}$ are the raising and lowering operators for the atom introduced in Eq. 4.2, $\hat{a}_{\mathbf{k}}^{\dagger}$ and $\hat{a}_{\mathbf{k}}$ are the creation and annihilation operators for photons in mode $\mathbf{k}$, and $g_{\mathbf{k}} = (w_{\mathbf{k}}/2\varepsilon_0\hbar V)^{1/2}\hat{\mathbf{e}}_{\mathbf{k}} \cdot \mathbf{d}_{eg}$ is the coupling constant for the electromagnetic field of mode $\mathbf{k}$ and the atomic dipole.

For an atom driven by a single-mode laser, the light field can be described by a coherent state $|\alpha\rangle$. Taking the expectation of the interaction Hamiltonian $\mathscr{H}_I$ with respect to the wavefunction of the coherent state, the creation and annihilation operators can be replaced with the eigenstates of Eq. 8.4 to give

$$\mathscr{H}_I = i\hbar g_{\mathbf{k}}\left\{\alpha\hat{\pi}^+(t)e^{i\mathbf{k}\cdot\mathbf{r}_A} - \alpha^*\hat{\pi}^+(t)e^{i\mathbf{k}\cdot\mathbf{r}_A}\right\}. \tag{C.2}$$

Comparing this to the interaction with a classical field from Eq. 4.3 yields $ig_{\mathbf{k}}\alpha = \Omega_p/2$ and $-ig_{\mathbf{k}}\alpha^* = \Omega_p/2$. Combining these and using the definition of the mean-photon number of a coherent state as $\bar{n} = |\alpha|^2$, the equivalence between an atom driven by a classical field with Rabi frequency Ω_p and a coherent-state is

J. D. Pritchard, *Cooperative Optical Non-Linearity in a Blockaded Rydberg Ensemble*,   155
Springer Theses, DOI: 10.1007/978-3-642-29712-0,
© Springer-Verlag Berlin Heidelberg 2012

$$\frac{\Omega_p^2}{4} \equiv g_\mathbf{k}^2 \bar{n}. \tag{C.3}$$

Thus for an atom displaced from the origin, the semiclassical coupling of Eq. 4.3 should be modified to include the position dependent phase-factors as follows,

$$V = \frac{\hbar\Omega}{2} \left(\hat{\pi}^+ e^{i\mathbf{k}\cdot\mathbf{r}_A} + \hat{\pi}^- e^{-i\mathbf{k}\cdot\mathbf{r}_A} \right). \tag{C.4}$$

For the simplified $g^{(2)}$ model in Sect. 8.4, a quantisation volume of $V = \pi\omega_0^2 c\Delta t$ is assumed for the probe laser, where w_0 is the $1/e^2$ beam radius. Taking the intensity of the laser as $I = 2P/\pi\omega_0^2$, Eq. C.3 reduces to

$$\bar{n} = \frac{2P\Delta t}{\hbar\omega}. \tag{C.5}$$

References

1. M.P. Gordon, P.R. Selvin, A microcontroller-based failsafe for single photon counting modules. Rev. Sci. Inst. **74**, 1150 (2003)
2. K.J. Weatherill, J.D. Pritchard, P.F. Griffin, U. Dammalapati, C.S. Adams, E. Riis, A versatile and reliably reusable ultrahigh vacuum viewport. Rev. Sci. Inst. **80**, 026105 (2009)
3. R. Loudon, *The Quantum Theory of Light*, 2nd edn. (Oxford University Press, Oxford, 1997)

Curriculum Vitae

Jonathan D. Pritchard

Education

2007–2011   Durham University, Ph.D. in Physics awarded April 2011

2003–2007   University of Nottingham, MSci Physics (1st class)

1997–2003   Sir Henry Floyd Grammar School, Buckinghamshire

Current Research

2011–Present   Post-doctoral research assistant in the photonics group at the University of Strathclyde, Scotland, developing an atomic interferometer using an inductively coupled ring trap.

Publications

- J. D. Pritchard, C. S. Adams and K. Mølmer, *Correlated Photon Emission from Multiatom Rydberg Dark States*, Phys. Rev. Lett **108**, 043601 (2012).
- M. Esmann, J. D. Pritchard, and C. Weiss, *Fractional photon-assisted tunnelling of ultra-cold atoms in periodically shaken double-well lattices*, Laser Phys. Lett. **9**, 160 (2012).
- S. Sevinçli et al., *Quantum interference in interacting three-level Rydberg gases: Coherent Population Trapping and Electromagnetically-Induced Transparency*, J. Phys. B **44**, 184018 (2011).
- J. D. Pritchard, A. Gauguet, K. J. Weatherill, and C. S. Adams, *Optical non-linearity in a dynamical Rydberg gas*, J. Phys. B **44**, 184019 (2011).

J. D. Pritchard, *Cooperative Optical Non-Linearity in a Blockaded Rydberg Ensemble*,   157
Springer Theses, DOI: 10.1007/978-3-642-29712-0,
© Springer-Verlag Berlin Heidelberg 2012

- M. Tanasittikosol, J. D. Pritchard, D. Maxwell, A. Gauguet, K. J. Weatherill, R. M. Potvliege, and C. S. Adams, *Microwave dressing of Rydberg dark states*, J. Phys. B **44**, 184020 (2011).
- J. D. Pritchard, D. Maxwell, A. Gauguet, K. J. Weatherill, M. P. A. Jones and C. S. Adams, *Cooperative atom-light interaction in a blockaded Rydberg ensemble*, Phys. Rev. Lett. **105**, 193603 (2010).
- K. J. Weatherill, J. D. Pritchard, P. F. Griffin, U. Dammalapati, C. S. Adams, and E. Riis, *A versatile and reliably re-usable ultra-high vacuum viewport*, Rev. Sci. Inst. **80**, 026105 (2009).
- R. P. Abel, A. K. Mohapatra, M. G. Bason, J. D. Pritchard, K. J. Weatherill, U. Raitzsch, and C. S. Adams, *Laser frequency stabilization to highly excited state transitions using electromagnetically induced transparency in a cascade system*, Appl. Phys. Lett. **94**, 071107 (2009).
- O. Kuras, J. D. Pritchard, P. I. Meldrum, J. E. Chambers, P. B. Wilkinson, R. D. Ogilvy and G. P. Wealthall, *Monitoring hydraulic processes with Automated time-Lapse Electrical Resistivity Tomography (ALERT)*, Comptes Rendus Geosciences **341**, 868 (2009).
- K. J. Weatherill, J. D. Pritchard, R. P. Abel, M. G. Bason, A. K. Mohapatra, and C. S. Adams, *Electromagnetically induced transparency of an interacting cold Rydberg ensemble*, J. Phys. B **41**, 201002 (2008).

Selected Presentations

- *Correlated Photon Emission from Multiatom Rydberg Dark States*. Poster presentation, Dynamics and simulation of Ultra-Cold Matter Workshop, Windsor, August 2011.
- *Light propagation through a strongly interacting Rydberg ensemble*. Invited talk, Cold Rydberg Gases and Ultracold Plasmas Workshop, Dresden, September 2010.
- *Cooperative Optical Non-linearity in an interacting Rydberg ensemble*. Poster presentation, 450th WE-Heraeus-Seminar, Bad Honnef, February 2010.
- *Light Interactions in Rydberg Ensembles*, Prize winning poster at QuAMP, Nottingham, September 2008.

Awards

- Department of Physics Thesis Prize of Durham University, 2011
- North Holland Research Physics Prize of Durham University, 2008
- Durham Doctoral Training Fellowship, 2007–2010
- Bill Moore Prize and Salmon Prize of the University of Nottingham, 2007